5th International Conference on Chemical Materials and Process (ICCMP 2019)

IOP Conference Series: Materials Science and Engineering
Volume 617

Bangkok, Thailand
1 – 3 May 2019

ISBN: 978-1-5108-9824-0
ISSN: 1757-8981

Printed from e-media with permission by:

Curran Associates, Inc.
57 Morehouse Lane
Red Hook, NY 12571

Some format issues inherent in the e-media version may also appear in this print version.

Institute of Physics
Dirac House, Temple Back
Bristol BS1 6BE UK

Phone: 44 1 17 929 7481
Fax: 44 1 17 920 0979

techtracking@iop.org

Additional copies of this publication are available from:

Curran Associates, Inc.
57 Morehouse Lane
Red Hook, NY 12571 USA
Phone: 845-758-0400
Fax: 845-758-2633
Email: curran@proceedings.com
Web: www.proceedings.com

Table of contents

Volume 617

5th International Conference on Chemical Materials and Process

1–3 May 2019, Chulalongkorn University, Bangkok, Thailand

Accepted papers received: 21 August 2019
Published online: 8 October 2019

Preface

Papers

Material Preparation and Performance Analysis

Material Science and Engineering

Chemical Application

Environmental Engineering

IOP Conf. Series: Materials Science and Engineering **617** (2019) 011001 doi:10.1088/1757-899X/617/1/011001

PREFACE

It is our great pleasure to welcome you to attend 2019 5th International Conference on Chemical Materials and Process (ICCMP 2019) which was held in Chulalongkorn University, Bangkok, Thailand during May 1-3. ICCMP 2019 is dedicated to issues related to Chemical Materials and Process.

The major goal and feature of the conference is to bring academic scientists, engineers, industry researchers together to exchange and share their experiences and research results, and discuss the practical challenges encountered and the solutions adopted. Professors from Canada, UK and Kuwait are invited to deliver keynote speeches regarding latest information in their respective expertise areas. It will be a golden opportunity for the students, researchers and engineers to interact with the experts and specialists to get their advice or consultation on chemical materials and process.

This proceeding present a selection from papers submitted to the conference from universities, research institutes and industries. All of the papers were subjected to peer-review by conference committee members and international reviewers. The papers selected depended on their quality and their relevancy to the conference. The volume tends to present to the readers the recent advances in the field of Multi-scale and/or Multi-disciplinary Approaches, Controlled release of the active ingredient, advanced materials processing, New materials & structured products, Environmental engineering & management etc.

We would like to appreciate all the authors who have contributed to this volume and also to the organizing committee, reviewers, speakers, chairpersons, sponsors and all the conference participants for their support to ICCMP 2019. We look forward to meeting you in ICCMP 2020!

Prof. Henry Hu
University of Windsor, Canada
July 12, 2019

Conference Committee

Conference General Co-Chairs

Prof. Paisan Kittisupakorn, Chulalongkorn University, Thailand
Prof. Henry Hu, University of Windsor, Canada

Program Co-Chairs

Prof. Jian Chen, Tsinghua University, China
Assoc. Prof. Zhigang Tang, Tsinghua University, China

Technical Committee

Prof. Yu Frank Yang, East Carolina University, USA
Prof. Mike Finnis, Imperial College London, United Kingdom
Prof. Gustaaf Schoukens, Ghent University, Department of Textiles, Belgium
Prof. Zizi I. Abdeen, Egyptian Petroleum Research Institute, Egypt
Prof. William Yu, Jilin University, China
Prof. Yebin Xu, Huazhong University of Science and Technology, China
Prof. Rongfei Zhou, Nanjing Tech University, China
Prof. Nour-Eddine Es-Safi, Mohammed V University in Rabat, Morocco
Prof. Parthasarathy Tigulla, Osmania University, India
Assoc. Prof. Jing Wang, University of South Florida, USA
Assoc. Prof. Dewei Chu, UNSW Sydney, Australia
Assoc. Prof. Yerdos Ongarbayev, Al-Farabi Kazakh National University, Kazakhstan
Assoc. Prof. Sujitra Wongkasemjit, Chulalongkorn University, Thailand
Assoc. Prof. Anton Dolzhenko, Monash University, Malaysia
Assoc. Prof. H. N. M. Ekramul Mahmud, University of Malaya, Kuala Lumpur, Malaysia
Assoc. Prof. Ravi Kumar, Dyal Singh College, India
Dr. Madhukumar R, RTES Degree college, Ranibennur Karnataka,India
Prof. Ashish Kumar, Lovely professional University, Punjab, India
Dr. Claudiu Constantin Manole, University "Politehnica" Bucharest, Romania
Dr. Chien-Hung Yeh, Department of Photonics, Feng Chia University, Taiwan
Dr. Jeyashelly Andas, Universiti Teknologi MARA, Malaysia

IOP Conf. Series: Materials Science and Engineering **617** (2019) 011003 doi:10.1088/1757-899X/617/1/011003

Peer review statement

All papers published in this volume of *IOP Conference Series: Materials Science and Engineering* have been peer reviewed through processes administered by the proceedings Editors. Reviews were conducted by expert referees to the professional and scientific standards expected of a proceedings journal published by IOP Publishing.

Synthesis and Characterization of Ordered Mesoporous Carbon CMK-3 with a High Loading Capacity of Ibuprofen and its Release Performance at Simulated Body Fluid

Maria Ulfa[1] and Didik Prasetyoko[2]

1 Study Program of Chemistry Education, Faculty of Teacher Training and Education, the March University, Jl. Ir. Sutami 36A, 57126 Surakarta, Central Java Indonesia
2 Department of Chemistry, Faculty of Mathematics and Natural Sciences, Institute of Technology Ten November, Jl. Keputih, Surabaya, East Java Indonesia
E-mail: ulfa.maria2015@gmail.com

Abstract. In this paper, ordered mesoporous carbon CMK-3 have been synthesized via hard templating methode using mesoporous silica as template followed by thermal carbonization. The resulting ordered mesoporous carbon sample have been loaded by ibuprofen as anti inflammantory drug model up to 75% (w/w). Characterization of textural and structural resulting ordered mesoporous carbon was examined by transmission electron microscopy, small angle X-ray scattering, Fourier transform infrared spectroscopy and nitrogen adsorption-desorption. The high loading ibuprofen effect in release performance have been reviewed and it appears that ibuprofen loading on ordered mesoporous carbon surface can be release up to 50% in simulated body fluid at pH 7.4 for 200 min. Result shows that high ibuprofen loading not only change the textural and structural material but also have great impact for release performance. Key Word: ordered mesoporous carbon, CMK-3, ibuprofen, high loading, release.

1. Introduction

Ordered mesoporous carbon (CMK-3) is a large porous adsorbent from the group nanomaterial which is considered as effective adsorbent, catalysis, energy storage and medical field [1]. The unique of CMK-3 is due to not only the availability of pores with a size range of 2-10 nm as access to large molecules but also regular pore structure, high surface area, inert and high thermal stability [2]. The pore size and surface area of ordered mesoporous carbon (CMK-3) can be good carrier for drug delivery material [3-4]. However, the using of CMK-3 for drug delivery system is high cost due to the complicated the synthesis process. To solve this problem, regeneration step is the best way to decrease the consume of CMK-3 but is not easy for oral drug consume. The alternative solution is how to control the CMK-3 pore in the releasing drug moleculewhich is need the information about the the change of structural and textural of material after drug loading process.

The popular anti inflammatory and antipyretic drug which is most used by all people around the world is Ibuprofen [5]. The pharmaceutical research was tried to increasing ibuprofen molecule loading for highly drug efficiencies by drug delivery system. Patient not need to consume drug gradually due to the high loading iburpofen in carrier material only if releasing process in long time period [6]. The various technique had been investigated to maximized the ibuprofen loading as adsorption and coagulation. Mesoporous carbon can be act as carrier of ibuprofen molecule in drug delivery system. However, nothing research was studied about the structural change of carrier material after loading

process. microporous activated carbon adsorbent had low efficiencies for decreasing ibuprofen molecule.
The control of loading process is required for the drug delivery applications [7-9]. Mesoporous carbons with uniform pores are expected to be carrier material for ibuprofen with hig loading capacities [10]. The high loading depend on pore size, surface area and surface chemistry of material [11-12]. However, in the best of our knowledge, comparation between material before and after laoding process and its imapct in realising process have been not studied before. Herein, we report a simple synthesis of CMK-3 via self-assembly route using hard templating method followed by loading profen using ibuprofen. The comaparation character of material before adn after loading process have been measured by characterization with X-ray diffraction (XRD), FTIR, N2 adsorption-desorption isotherms and transmission electron microscopy (TEM). Release capacity of CMK-3 have been observed by simulated body fluid to dissolute ibuprofen from CMK-3.

2. Experiment

2.1 Material

The materials used in this study tetraetylortosilicate (Merck, 98%), sulfuric acid, hydrochloric acid, n-hexane, NaOH (Merck, 99%), Pluronic P123, HF, hydrochloric acid, ethanol, methanol, acetone and aquabides, Ibuprofen (Merck, 98%).

2.2 Synthesis CMK-3

The mesopore carbon synthesis begins by SBA-15 synthesis according to the procedure elsewhere [14] . For CMK-3 synthesis was firstly started by mixing sucrose powder about 1.04 g in a 0.2 g mixture of concentrated sulfuric acid in 100 g air while stirring to produce a homogeneous sucrose solution. Then 1 g of SBA-15 was added to the sucrose solution and added homogeneously for 2 hours. Mixing until homogeneous. So that carbon is infiltrated into the mold and coating the mold walls. The bone mixture is then heated in an oven at 100 ° C for 7 hours then the temperature is raised to 160 ° C for 7 hours to dehydrate. After the sucrose / SBA-15 composite contains a period of partial carbonized organic, then added sucrose solution containing 0.92 g of sucrose into a mixture of 0.08 g sulfuric acid and 100 g of water. sucrose / SBA-15 composites were then heated again at 100 ° C for 6 hours followed by a temperature of 160 ° C for 6 hours into black powder.carbonization begins with carbonization of samples temperature of 30-300 ° C. The holding time is carried out for 1 hour at 300 ° C. Furthermore, carbonization is continued in the range of 300-600 ° C and is held for 1 hour at 600 ° C. Furthermore, the black powder is carbonized by heating 900 ° C for 3 hours. The entire carbonization is carried out at a heating rate of 5 ° C / min under the N2 gas flow. The carbonized sample is then washed with 1M NaOH (in ethanol and water 50:50) temperature 70 ° C to release the silica mold. In the next synthesis washing is done with HF with a concentration of 10-40% for 2-24 hours. Washing is done 3 times to make sure the silica is released all, followed by washing using ethanol. The resulting sample is then dried at 110 ° C and stored in the desiccator. The resulting sample was labelled as CMK-3 before loading

2.3 Ibuprofen loading-release

Ibuprofen-loaded CMK-3 was prepared as follows: 0.075 g of Ibuprofen was dissolved in 20 mL of ethanol, and 0.100 g of CMK-3 powder was then added under constant stirring for 24 h at room temperature. The solid part of the dispersion was separated using centrifugation and dried overnight at room temperature. The resulting sample was labelled as CMK-3 after loading. The drug release experiments were performed by immersing the drug loading samples into simulated body fluid (pH 7.4 at 37 °C). Several sample were tested in each 10 min and examined by UV–Vis spectroscopy.

3. Result and discussion

TEM results of CMK-3 before loading and CMK-3 after loading mesoporous carbon samples are shown in Figure 1. TEM images show the presence of uniform pores on CMK-3 before loading and CMK-3 after loading sizes of 4.5 and 3.8 nm seen from the vertical and horizontal directions.

Uniformity of the pore size was confirmed by the particle size histogram processed from TEM images. The particle size distribution of CMK-3 before loading samples shows the dominance of uniform particles measuring 1-5 nm. Meanwhile, the particle size distribution at CMK-3 after loading shows the dominance of uniform particles in the range of 5-10 nm. This is not much different from the pore size distribution resulting from the calculation of nitrogen adsorption-desorption data in the CMK-3 before loading and CMK-3 after loading samples which were 4.06 nm and 8.5 nm respectively. This fact reinforces the conclusion that the results of the CMK-3 before loading and CMK-3 after loading sample pore measurements are not much different despite using 2 different instruments.

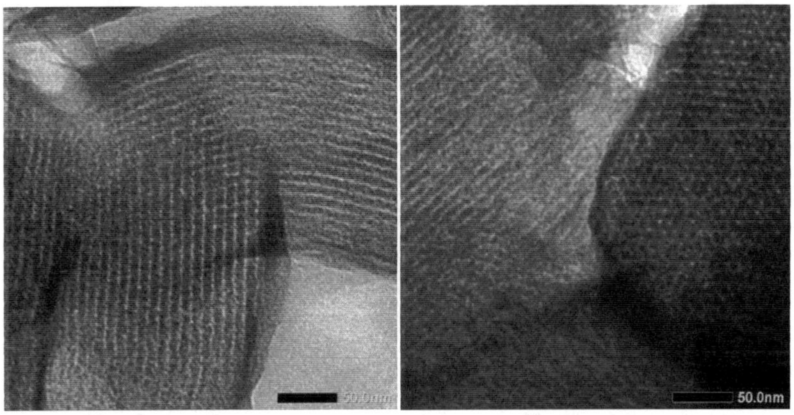

a. b.

Figure 1. Ilustration of mesoporous carbon CMK a. before and b. after iburprofen loading

Figure 1 shows CMK-3 before loading mesoporous carbon samples which is imply regular hexagonal structures with p6mm group space based on the pattern of black dots shown by FFT. The TEM image pattern shows that CMK-3 have a honeycomb structure as represents a hexagonal pore arrangement. CMK-3 morphology of parallel points representing the arrangement of carbon bars when observed from the horizontal direction. Figure 1 also shows a small number of irregular stacks (stacking faults) of CMK-3 after ibuprofen loading due to some skeletons experiencing little damage came from ibuprofen desposition in inner pore at high loading up to 75% w/w. The TEM observations on CMK-3 before loading provide the conclusion that mesoporous carbon has the same morphology as CMK-3 after loading silica which is could be said that not only nothing significant big destruction after ibuprofen loaded onto CMK-3 but also this implies the stability of CMK-3.

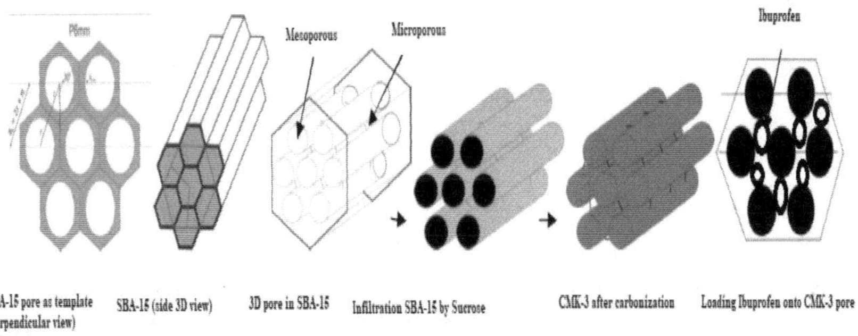

Figure 2. Illustration of mesoporous carbon CMK during synthesis and ibuprofen loading.

Figure 2 is an illustration that represents the SBA-15 replication process to become mesoporous carbon CMK-3. In the illustration, SBA has a pore structure arrangement that forms a hexagonal structure which is have small micropores or mesopores on the wall as bridges connecting the main channels of the mesoporous tube. If the carbon precursor fills the entire inner chamber of the SBA-15, a carbon frame will form with a solid stem. This process is enhanced by the release of the silica skeleton. When the silica wall has been released, the silica wall changes into an empty carbon pore channel. This means that the actual carbon pore is the former position of the released silica wall which is in next application has been used by ibuprofen as accesible pore during loading process. The ibuprofen have been loadedonto CMK-3 by fills larger the meso-empty carbon than micro-empty channel. It is logic because the ibuprofen size in more small than the mesoporous space in CMK-3. The high loading of ibuprofen in our prediction would be have great impact on surface chemistry of CMK-3 as FTIR observation below.

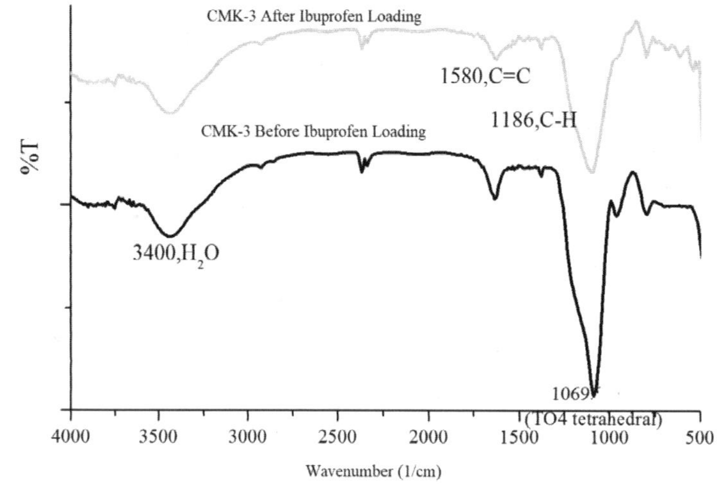

Figure 3. Spectra of FTIR on CMK-3 sample.

FTIR spectra in all CMK-3 samples showed wide absorption peaks at 3600-3100 cm^{-1} wave number not only indicates the presence of a number of hydrogen bonds but also describes the amount of water absorbed on the surface of the sample. The absorption band in the range of 1300-1000 cm^{-1} shows the vibration of the CO bond strain presented in the following functional groups: anhydride (980-1300 cm^{-1}), ester (1150-1080 cm^{-1}), lactone (1160-1370 cm^{-1}), alcohol (1280-1050 cm^{-1}), phenol (1220-1000 cm^{-1}), and ether (1300-1 000 cm^{-1}) [12]. The absorption area at the wavelength range 2000-1500 cm^{-1} with small intensity shows the characteristic character of double bond vibration. In the range of these wavelengths, there are several vibrations observed, including vibrations C = C (1680-1620 cm^{-1}), aromatic rings (1600-1585 cm^{-1}), C = O from carboxylates (1760-1665 cm^{-1}) , carbonyls (1600-1590 cm^{-1}), quinones (1680-1550 cm^{-1}), lactone groups (1790-1675 cm^{-1}), and anhydrides (1880-1740 cm^{-1}) [13]. The vibration of the CH3 and CH2 groups appears in the range of 3000-2800 cm^{-1}. Nothing significant differences wa observed but the intensity of CMK-3 after loading is slighty lower than before loading due to the ibuprofen was filled the CMK-3 pore.

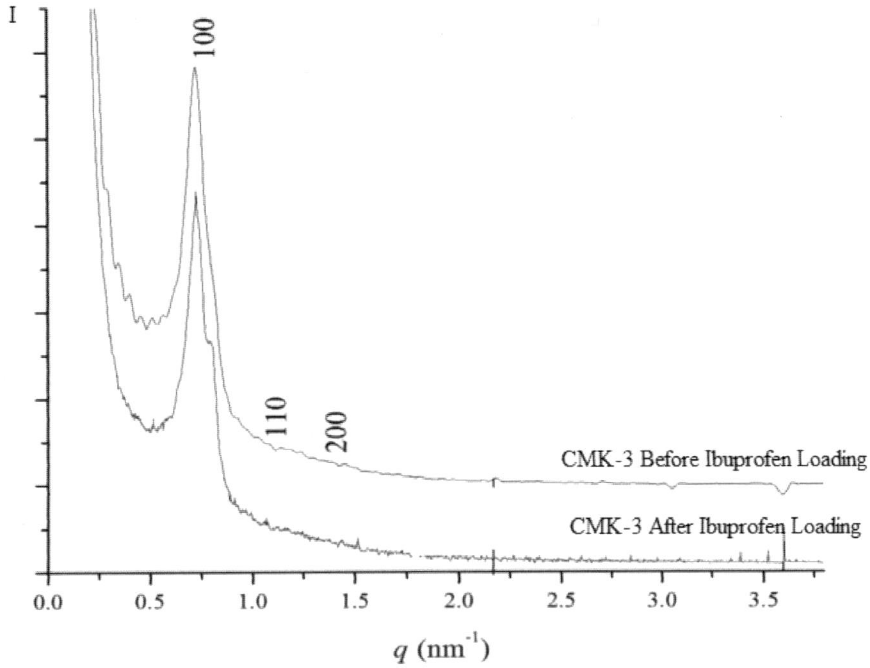

Figure 4. SAXS of CMK-3 before and after ibuprofen loading.

In this study, observations with the scattering results of SAXS were carried out on sample CMK-3 before and after ibuprofen loading. Figure 4 both samples show peaks (100), (110) and (200) which describe regular two-dimensional hexagonal mesopore arrangements. The intensity of the q values at the top (100), (110) and (200) positions observed are 0.72; 0.79; and 1.09 where the q peak at CMK-3 after ibuprofen loading is lower than CMK-3 after ibuprofen loading as the effect of the former of ibuprofen sites. The SAXS scattering results in all samples showed peaks with sufficiently strong intensity at (100) and (200) which indicated the regularity of the mesoporous structures formed. The size of the pore CMK-3 before and after ibuprofen loading is 3.39 and 3.16 nm. The pore of CMK-3 before ibuprofen loading is greater than after which indicates that small part of CMK-3 frame work may desctructed by ibuprofen after loading. This phenomenon is consistent with the explanation of next nitrogen adsorption-desorption isotherm data and TEM observations. Overall, CMK-3 before not significantly different from CMK-3 before ibuprofen loading, so it can be concluded that the ibuprofen loading at 75% w/w remaining the the CMK-3 structure

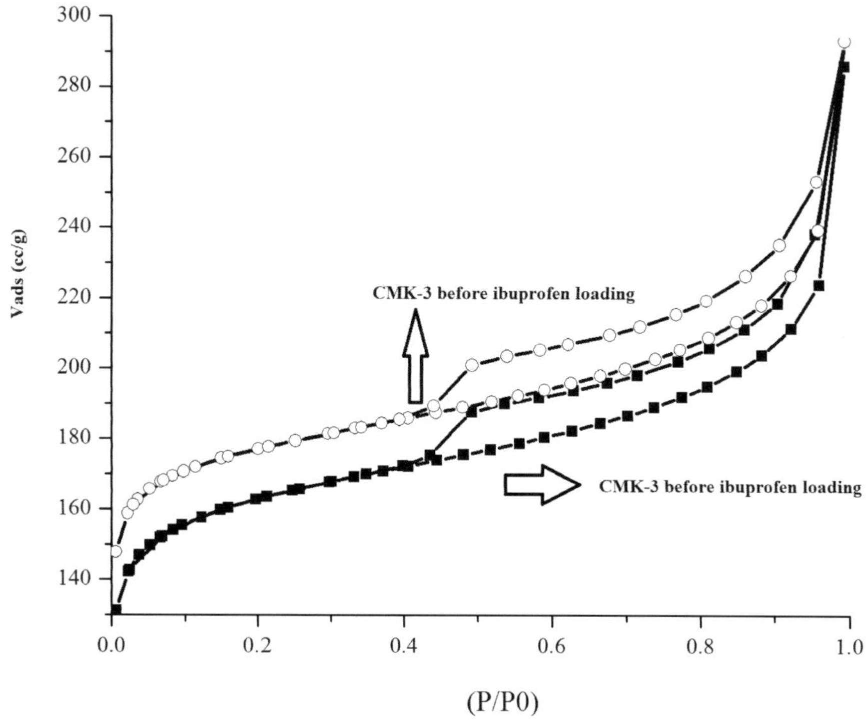

Figure 5. Isotherm-adsorption-desorption nitrogen CMK-3 before and after ibuprofen loading.

Figure 5 shows the nitrogen adsorption isotherm in the CMK-3 sample.The successful synthesis of mesoporous carbon can also be seen by observing the inflection point of the nitrogen adsorption isotherm in the CMK-3 sample. The position of P / Po at the inflection point is a function of pore diameter due to the presence of surface adsorption procedures along with the formation of mono and multilayer [14-15]. Infection of the isoterm of mesoporous CMK-3 carbon samples at P / Po 0.0-1.0 is divided into 3 areas. First, the inflection point at P / Po <0.5 where nitrogen fills a small portion of the pore volume in small pores representing the formation of a monolayer on micropore carbon. Furthermore, most of the porous space with a larger size is filled with nitrogen along with the increase in pressure (P / Po 0.4-0.8) which reflects the contribution of mesoporous to carbon during adsorption. In this area there is a sharp inflection on carbon CMK-3 isotherm as more nitrogen enters is estimated due to nitrogen capillary condensation in the mesopore. According to some researchers, this phenomenon is a character of the pore uniformity of a material [16-17]. Sharp inflection in the pressure range is a sign that there is an interconnection of the pores which gives another meaning that in the pore tissue there is a primary pore and there are secondary pores (generally smaller). In some studies, smaller pore on CMK-3 framework have been acted as pore neck or pore bridges that connecting each mesorod carbon which is impact to stability building block [18-19]. The area after the capillary condensation stage appears as a displaying plain at higher relative pressures that are expected to occur due to the formation of multilayer nitrogen adsorption on the external surface and empty slits of CMK-3 samples. Shallow slope in this area indicates that the external surface area in the CMK-3 sample is very small. The desorption branch almost did not coincide with the adsorption branch which revealed that nitrogen adsorption occurred irreversibly in mesopore. This is thought to be due to a pause when capillary condensation occurs. Finally, at high pressure (P / Po> 0.8) an inflection point occurs in the vertical direction which is estimated to be the area of entry of nitrogen in macropore carbon. The isoterm CMK-3 after loading was lower than before loading due to the pore filled by ibuprofen. This loading process have big impact to the CMK-3 pore because inner pore was covered by ibuprofen molecule so nitrogen can not fill the pore.

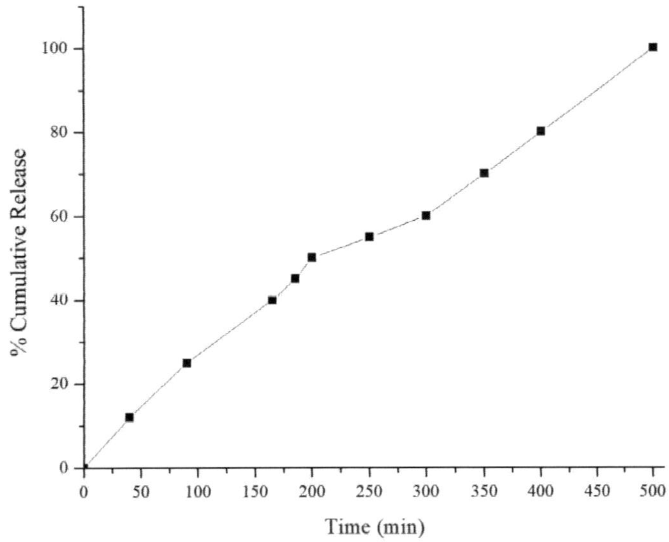

Figure 6. Release trend of ibuprofen using CMK-3 after high loading process.

Figure 6 shows the ibuprofen release profile with CMK-3 for 500. The ibuprofen release profile with CMK-3 is found to reach 50% at first 200 min and it is fully release at 500 min. This phenomenon indicating that ibuprofen molecule adsorpt onto both of meso and microporous part of CMK-3. The slow release profile in our prediction have been affected by ibuprofen deposition in inner pore of CMK-3. The high deposition may affected by hig loading ibupfofen and numerous of mesopore space which is available for ibuprofen as big molecule. The second prediction is ibuprofen have been trapped not only by functional group of CMK-3 surface but also agglomertion of ibuprofen molecule in depth surface. a reduction of% dissolution of original ibuprofen due to the formation of agglomerates. This is in agreement with the nitrogen adsorption data, morphology of material by TEM, structure profile by XRD and implies the presence of free hydroxyl group as active sites on the surface of CMK-3 as evidenced by FTIR spectra. Over all, The textural and structural of CMK-3 before ibuprofen loading was very similar to CMK-3 before ibuprofen loading which is imply the stabililty of CMK-3 for regeneration carrier material in drug delivery system

4. Conclusion

Ordered mesoporous carbon CMK-3 was succesfully constructed by via hard templating methode using mesoporous silica as template followed by thermal carbonization. The result show that at 75% w/w loading capacity of ordered mesoporous carbon by ibuprofen have not dramatically destructed the structure. The characterization of ordered mesoporous carbon show that any small sringkage part of morphology due to the ibuprofen removing process. Over all, surface chemistry of CMK-3 after and before ibuprofen loading is slighty different in surface area, scattering and functional group intensity but not significant in structure. The ibuprofen release result show that ordered mesoporous carbon can releasing ibuprofen up to 50% in simulated body fluid at pH 7.4 for 200 min. It is good information that CMK-3 in potential material for drug delivery system in the future.

5. References

[1] G. Chandrasekar, W. J. Son, and W. S. Ahn, *J. Porous Mater.*, 2009.
[2] H. Zhou, S. Zhu, I. Honma, and K. Seki, " *Chem. Phys. Lett.*, 2004.

[3] N. Farzin Nejad, E. Shams, M. K. Amini, and J. C. Bennett, " *Microporous Mesoporous Mater.*, 2013.
[4] M. Ulfa, R. M. Hasanah, and D. Prasetyoko,*Proc. Int. Conf. on Physics (AIP)* vol. 020001, 2018.
[5] S. Mondal, K. Aikat, and G. Halder, *Ecol. Eng.*, vol. 92, pp. 158–172, 2016.
[6] F. Rehman, P. L. O. Volpe, and C. Airoldi, " *Colloids Surfaces B Biointerfaces*, vol. 119, pp. 82–89, 2014.
[7] J. Kurczewska, D. Lewandowski, A. Olejnik, and G. Schroeder, " *Int. J. Pharm.*, vol. 472, no. 1–2, pp. 248–250, 2014.
[8] S. Jabeen, M. Maswal, O. A. Chat, G. M. Rather, and A. A. Dar, *Colloids Surfaces B Biointerfaces*, vol. 139, pp. 211–218, 2016.
[9] L. Zhang, H. Cheng, C. Zheng, F. Dong, S. Man, and Y. Dai, *J. Drug Deliv. Sci. Technol.*, vol. 31, pp. 101–107, 2016.
[10] Y. Zhang, J. Chen, G. Zhang, J. Lu, H. Yan, and K. Liu, *React. Funct. Polym.*, vol. 72, no. 6, pp. 359–364, 2012.
[11] L. Qu, Q. Tony, J. A. Denman, P. J. Stewart, K. P. Hapgood, and D. A. V Morton, *European Journal of Pharmaceutical Sciences* vol. 78, pp. 264–272, 2015.
[12] A. Guglielmelli, B. Rizzuti, and R. Guzzi, *European Journal of Pharmaceutical Sciences* vol. 112, no. November 2017, pp. 122–131, 2018.
[13] X. Jin, Q. Wang, J. Sun, H. Panezai, S. Bai, and X. Wu, " *Microporous Mesoporous Mater.*, vol. 254, pp. 77–85, 2017.
[14] D. Tan *et al.*, *Appl. Clay Sci.*, vol. 96, pp. 50–55, 2014.
[15] P. a. Bazuła, A.-H. Lu, J.-J. Nitz, and F. Schüth, *Microporous Mesoporous Mater.*, vol. 108, no. 1–3, pp. 266–275, 2008.
[16] D. J. Kim, H. I. Lee, J. E. Yie, S. J. Kim, and J. M. Kim, *Carbon N. Y.*, vol. 43, no. 9, pp. 1868–1873, 2005.
[17] A. Vinu, M. Miyahara, and K. Ariga, G. Liu, S. Zheng, D. Yin, Z. Xu, J. Fan, F. Jiang, *J. Colloid Interface Sci.*, vol. 302, no. 1, pp. 47–53, 2006.
[18] L.A Solovyov, A.N Shmakov, V.I Zaikovskii, S.H Joo, R Ryoo., *Carbon*, Vol 40, Issue 13, 2477-2481, 2002
[19] W. Schmidt, *Microporous Mesoporous Mater.*, Vol 117, Issues 1–2,372-379, 2009

Acknowledgements
The authors would like to thank Directorat General of High Education (DIKTI), Indonesia, for the financial support through a Post Doctoral Research 2019 scheme grant with number contract of grant 718/UN27.21/PN/2019 (DIKTI Number contract 092/SP2H/LT/DRPM/2019)

IOP Conf. Series: Materials Science and Engineering **617** (2019) 012002 doi:10.1088/1757-899X/617/1/012002

Synthesis and Characterization of Pectin-Chitosan as Candidate Materials for Slow Release System

Budi Hastuti[1] and Saptono Hadi[2]

1 Department of Chemistry Education, Faculty of Teacher Training and Education, Universitas Sebelas Maret
2 Department of Pharmacy, Faculty of Mathematics and Natural Sciences, Universitas Sebelas Maret
E-mail: budihastuti@staff.uns.ac.id[1]

Abstract. The polyelectrolyte complex (PEC) is a complex formed between opposite charge particles which have recently attracted attention for their potential for drug delivery applications. In this study, PEC from pectin (anionic species) and chitosan (cationic species) was studied. Pectin and chitosan solutions in 2% v/v acetic acid were mixed in the ratio pectin : chitosan of 1:1. The difference in characteristics of PEC pectin-chitosan (Pec-Chi) with its original polymer is indicated by functional group analysis using FTIR. Morphology characterization of films were investigated by XRD. The results showed that Pec-Chi adsorbent had better stability and resistance to the acidic environment. Characterization using FTIR showed that the adsorbent has functional groups of amines, carboxyl and hydrogen. Furthermore, XRD data show that the adsorbent was a not crystalline.

1. Introduction

The use of natural polymers in the design of drug delivery system has received attention in recent years, due to their excellent bioavailability and biodegradability. As the oral route is the most common route for drug administration, the use of polymers in controlling the release of drug has become the most important tool in the formulation where the drug is incorporated in the polymers. Related to this, understanding factors that influence drug release kinetics during the process of adsorption, distribution, metabolism and excretion (ADME) is very important, in order to produce better pharmacological action [1]. Furthermore, in vitro drug dissolution studies become a very important part in the development of drug technology because it is a beginning in examining the effectiveness of drug release.

Research on drug release from its matrices such as natural polysaccharides as matrices in drug delivery systems have been extensively investigated in their application [2]. Chitosan and its derivatives is an example of a polysaccharide which have been studied in drug delivery systems in various forms [3]. Chitosan is biocompatible, biodegradable and non-toxic so that it can be used as a matrix in controlled curcumin release systems [4]. Chitosan is a cationic (positively charged) polymer derived from poly-N-acetyl-D-glucosamine which is biodegradable, biocompatible, non-toxic and is often used extensively in the biomedical field [5]. The positive charge of chitosan is very important in the drug delivery system because it plays a role in interactions with the drugs it delivers. The interaction of chitosan with drugs can increase solubility, stability and bioavalaibility of encapsulated drugs in it [6].

The use of chitosan as an ingredient in controlled drug release matrices has the disadvantage that the mechanical strength of chitosan membranes is very low that the membrane is fragile. In addition, chitosan has low hydrophilicity properties so it is necessary to modify chitosan physically and chemically to increase the mechanical strength and hydrophilicity of the membrane. One of them is by adding anionic polysaccharide which has a higher hydrophility than chitosan. Pectin can improve mechanical properties and increase the hydrofility of chitosan membranes. Pectin is soluble in water because it has a carboxyl group [7]. Pectin is biodegradable and non-toxic so it is widely used in the pharmaceutical industry for drug delivery in the form of tablets, gels and film membranes.
Chen et al. (2010) make chitosan-pectin composites to increase membrane strength so that they can control hydrophilicity and disintegration [8]. The research proves that chitosan-pectin composite membrane shows the right balance between the hydrophilic group and hydrophob which is the nature of a film. Chitosan pectin membranes have greater hydrophilicity and water absorption compared to pure membranes (chitosan or pectin only). Therefore, chitosan-pectin composite membranes is expected as biocompatible, biodegradable and non-toxic material, that thus suitable to be applied as drug release agent.

2. Materials and method

2.1. Material
Chitosan purchased from CV Ocean Fresh Bandung, pectin from apple waste purchased from UD Organic Yogyakarta, HCl (E-Merck), isopropanol (E-Merck), chloroacetic acid (E-Merck), NaOH (E-Merck), acetic acid (E-Merck), and aquadest.

2.2. Preparation of the membrane.

2.2.1 Synthesis of carboxymethyl chitosan (CMC) by grafting chitosan with chloroacetate. 3 g chitosan powder was dissolved in 80 mL of isopropanol then was added with drops of 40% NaOH as much as 28 mL. Chitosan and isopropanol were stirred for 30 minutes. Then chloroacetic acid is added with the mole ratio of chloroacetic acid : chitosan is 10 : 1. The mixture is filtered and washed with aquabides and ethanol. The filtered chitosan deposit is dried and ovened at 60° C for 3 hours. Functional group characterization was performed using FTIR spectroscopy.

2.2.2 Synthesis of Pectin-CMC adsorbent. 0.2 g CMC powder was dissolved in 10 mL of 5% acetic acid. 0.2 g pectin powder was dissolved in 10 mL of 5% acetic acid. Both solutions were mixed, stirred for 2 hours then printed in petridish and dried. After drying the membrane is removed from the mold by adding 10 mL 1 M NaOH. The membrane that has been separated from the mold is washed with distilled water and then dried. Functional group characterization was performed using FTIR spectroscopy, characterization using XRD and surface analysis using SEM.

3. Results and discussion
The synthesis of PEC pectin-carboxymethyl chitosan membrane as an intermediate delivery material for this drug was initially carried out by grafting chitosan with a monocloroacetic group so that carboxymethyl chitosan was formed. After that it is composited with a pectin group, so that PEC Pectin-Carboxymethyl chitosan (CMC) membrane will form, which is a membrane that is more stable and more soluble in water. Thus PEC Pectin-Chitosan has the potential to be used as a slow release material for drugs.
Chitosan has low hydrophilicity properties so it is necessary to modify chitosan physically and chemically to increase the mechanical strength and hydrophilicity of the membrane. The addition of pectin which is anionic polysaccharide which has a higher hydrophility than chitosan can improve the mechanical properties and increase the hydrofility of the chitosan membrane. Thus the chitosan-pectin

IOP Conf. Series: Materials Science and Engineering **617** (2019) 012002 doi:10.1088/1757-899X/617/1/012002

composite membrane increases the hydrophilicity and absorption of water in the composite membrane, so that it has the potential as a drug disintegration.

3.*1*. *Functional group analysis uses FTIR*

In order to prove that the –COOH group is the dominant active group in pectin, the $-NH_2$ group is the active group that is dominant in chitosan while the active group which is dominant in CMC is the $-NH_2$ and –COOH groups. Furthermore, PEC Pectin-CMC membrane was tested for functional groups using an infrared spectrophotometer. FTIR spectra are presented in Figure 1.

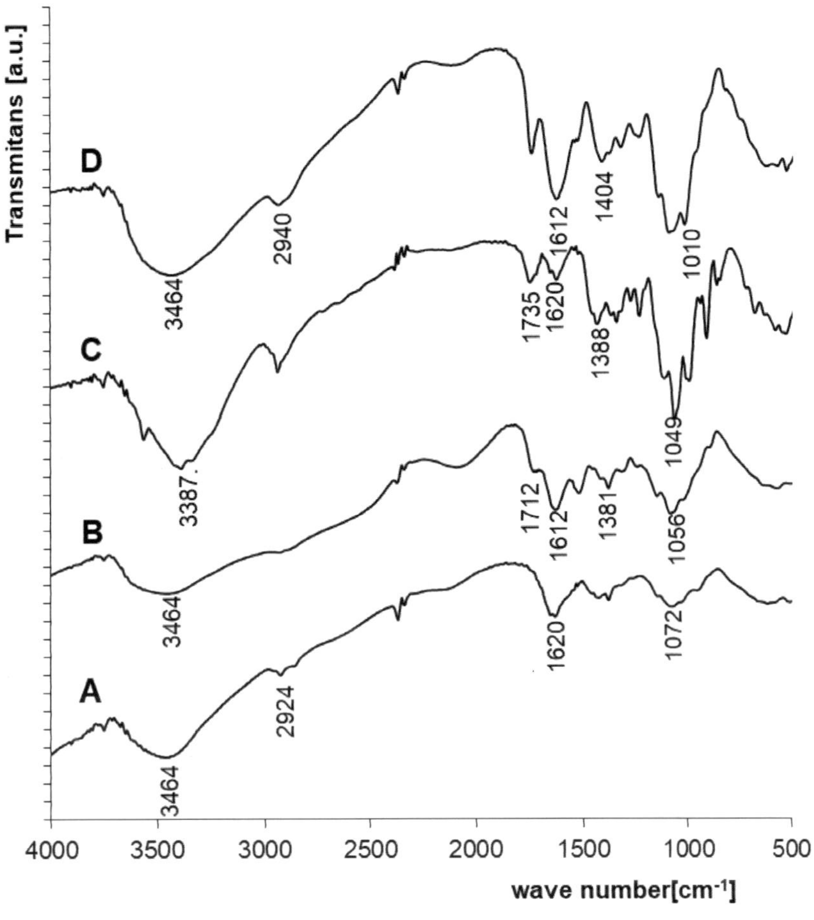

Figure 1. FTIR spectra of A. chitosan, B. CMC, C. Pectin and D. Pectin-CMC compounds.

In Figure 1, it appears that in chitosan spectra, CMC has a wave number of 1597 cm^{-1} as –NH absorption (amine) and 3464 cm^{-1} as –OH absorption. In the pectin spectra there is a wavenumber 1627 cm^{-1} as –COOH absorption and at wavenumber 1751 cm^{-1} as C=O ester. Pectin-CMC spectra showwave number 1420 cm^{-1} as vibration absorption peak –NH (amine), 1635 cm^{-1} as –COOH absorption, and 1735 cm^{-1} represents stretching vibration absorption –C=O ester, 2940 cm^{-1} is stretching vibration absorption of the $O\text{-}CH_3$, and 3464 cm^{-1} groups is the absorption of –OH groups.

3.2. *Analysis of crystallinity with XRD*

Analysis with XRD spectrometry was intended to determine the crystallinity of chitosan, pectin, chitosan after chloroacetate and CMC grafting after being added to pectin. The diffractogram of chitosan, CMC, pectin and Pectin-CMC is shown in Figure 2.

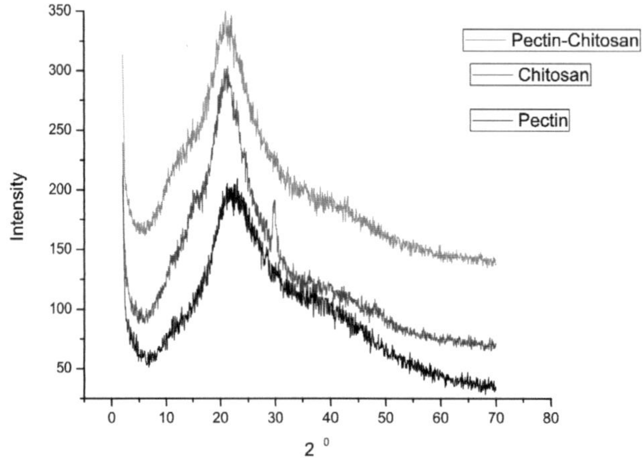

Figure 2. Diffractogram of XRD from chitosan, CMC, Pectin and Pectin-CMC.

The diffractogram of chitosan, CMC and Pectin-CMC showed a widening peak at $2\theta = 10°$, and $2\theta = 20°$. Peak $2\theta = 10°$, and $2\theta = 20°$ in CMC decreased compared to the initial chitosan. This shows that there is a decrease in the degree of crystallinity due to the grafting process. The transplantation process changes the side groups by replacing hydrogen atoms with acetic acid groups so that the hydrogen bonds are weakened. Decreased crystallinity of chitosan by chloroacetate grafting was also shown by Abreu and Campana-Filho (2005). The process of combining CMC compounds with Pectin caused a change in diffractogram, non crystalline and pectin CMC which was crystalline after being combined to form semi-crystalline Pectin-CMC compounds. The decrease in crystallinity is also shown by Pectin-CMC compounds, this indicates that PEC Pectin-CMC to absorb moisture in it is useful for preparation of hydrogels.

4. Conclusion

The PEC Pectin-CMC membrane synthesized in this study can be applied as a drug release matrix. From the FTIR test results showed that the PEC Pectin-CMC membrane had carbonyl group (C=O), amine group ($-NH_4$) and hydroxyl group (–OH). The XRD test showed that the membrane showed a decrease in crystallinity, which showed that the membrane could be applied as a matrix of slow release drug.

5. Reference

[1] Singhvi, G. and Singh, M., 2011, Review: In-Vitro Drug Release Characterization Models, *Int. J. Pharm. Res.*, 2, 77-84.
[2] Sriamornsak, Pornsark., 2003, Chemistry of Pectin and Its Pharmaceutical Uses: A Review, international Journal, Vol. 3, Silpakorn University
[3] Luo, Y., Teng, Z., Wang, X., and Wang, Q., 2012, Development of Carboxymethyl Chitosan Hydrogel Beads in Alcohol–Aqueous Binary Solvent for Nutrient Delivery Applications, *Food Hydrocol.,* 31, 332–339
[4] Luo, Y. and Wang, Q., 2014, Recent Development of Chitosan–Based Polyelectrolyte Complexes with Natural Polysaccharides for Drug Delivery, *Int. J. Bio. Macro.*, 64, 353– 367.
[5] Illum, L., 2007, Nanoparticulate Systems for Nasal Delivery of Drugs: A Real Improvement Over Simple Systems, *J. Pharm. Sci.*, 96, 473-483.
[6] Morris, E.R., Rees, D.A., Thom, D. dan Boyd, J., 1978, Chiroptical and Stoichiometric Evidence of a Specific, Primary Dimerization Process in Alginate Gelation, *Carbohydr. Res.*, 66, 145-154

[7] Hastuti, B., Siswanta, D., Mudasir, Triyono., 2017, Kinetics and Isotherm Studies of Pb(II) Imprinted Carboxymethyl Chitosan Pectin PEGDE, Jurnal Teknologi, 79:6, 141-147
[8] Chen, P., Kuo, T.Y., Kuo, J.R., Tseng, Y.P., Wang, D.M, Lai, J.Y., Hsieh, H.J. 2010. Novel Chitosan-Pectin Composite Membranes with Enhanced Strength, Hydrophilicity and Controllable Disintegration, Carbohydr. Polym. 82, 1236-1242.

IOP Conf. Series: Materials Science and Engineering **617** (2019) 012003 doi:10.1088/1757-899X/617/1/012003

In-situ synthesis of $CdS@TiO_2$ Nanoparticles-nanobelt Heterojunctions with high photocatalytic property

YM Wang[1], G J Du[2, 3] and M L Qi[1]
1 School of transportation and civil engineering, Shandong Jiaotong University, Jinan 250357, China
2 College of Chemistry and Molecular Engineering, Peking University, Beijing 100871, China
3 Fasten Group Ltd., Jiangyin, Jiangsu 214434, China
Email: ymwangsdu@139.com

Abstract. In this study, CdS nanoparticles were in-situ deposited on the surface of TiO_2 nanorods through the liquid phase synthesis method, and the $CdS@TiO_2$ Nanoparticles-nanobelt heterojunctions were obtained. The nanobelts are 50-200nm in width and several tens of micrometers in length, while the size of CdS nanoparticles is about 10 nm. The photocatalytic properties of the products were also studied and the results showed that the photocatalytic activity of CdS@ TiO_2 heterostructure is improved greatly compared with that of the common nanobelts. This is attributed to the different band energy structure of the two semiconductors, facilitating efficient charge separation and migration.

1. Introduction

Semiconductor Photocatalytic technology, which is one of the effective ways to solve the energy and environment problems fundamentally, has brought great economic, environmental and social benefits in the field of environmental management and solar energy utilization [1]. However, there are still some problems in photocatalytic technology, such as low quantum yield, high composite efficiency of photocarriers and low utilization rate of sunlight. Heterojunctions, z-scheme photocatalyst, noble metal deposition and dye sensitization can reduce the recombination efficiency of photogenerated electrons/holes by changing transport route or capturing photogenic carriers [2-5].

Titanium dioxide (TiO_2) and cadmium sulfide (CdS) are both important multifunctional common semiconductor materials. If CdS is assembled on the surface of TiO_2 nanobelt, schottky junction will be formed on the surface, and these heterogeneous junctions will play a key role in the photocatalytic performance. Different composite structures of CdS and TiO_2 have been prepared by different methods and the application prospect has been discussed in literatures, such as CdS nanorod-coated TiO_2 nanowires [6], TiO_2 nanostructures with CdS quantum Dots[7-9], TiO_2 nanotube array-graphene-CdS quantum dots composite film[10], CdS/reduced graphene oxide/TiO_2 core–shell nanostructure [11] and C_{60}-decorated CdS/TiO_2 mesoporous architectures [12]. However, the heterostructure of CdS nanoparticles and TiO_2 nanobelts has rarely been reported.

In this study, CdS nanoparticles were assembled in situ on the surface of TiO_2 nanobelts, which was obtained by alkali-assisted hydrothermal and subsequent treatment method. The as-prepared samples were characterized by scanning electron microscope (SEM), transmission electron microscope (TEM), X-ray diffraction (XRD) and Ultraviolet visible (UV-vis) absorption spectrum. The photocatalytic performance was evaluated by degradation of methyl orange solution and the photocatalytic mechanism was also discussed.

2. Experimental

TiO_2 nanobelts was prepared by the alkali-assisted hydrothermal and subsequent heat treatment method with the heat treatment temperature of 600°C for 1 h, and the detailed preparation process described in the literature [13]. The CdS@TiO_2 nanoparticle-nanobelt heterostructure was synthesized in situ by liquid phase method [14]. The typical process is as follows: 80 mg TiO_2 nanosbelts was ultrasonic dispersed in 109 mL deionized water, and then the suspension is heated by the water bath along with magnetic stirring. The temperature of the reaction mixture was monitored using a thermometer. The 20 mL mixture solution containing 0.055 g $CdCl_2$ and 0.16 g NH_4Cl was added slowly. When the temperature reached 80 °C, 15 mL thiourea solution, which contains 0.233 g thiourea, was added to mixture solution and then the temperature of reaction solution decrease. When the temperature of reaction solution went back to 80 °C, 6 mL 28~30wt% NH_4OH solution was added quickly to begin the reaction for 2 min. Finally, after washing thoroughly with water and drying at 60°C, CdS@TiO_2 nanoparticles-nanoribbon heterojunction could be obtained.

X-ray Powder diffraction (XRD) patterns were recorded with Cu-Ka (λ=0.15418 nm) radiation and a scan rate of 0.1° s^{-1} at 2θ ranging from 20° to 60°. The morphologies of the samples were observed by a field emission scanning electron microscope (SEM), transmission electron microscope (TEM) and high-resolution transmission electron microscope (HRTEM). Ultraviolet visible (UV-vis) absorption spectrum were obtained by using a UV–vis spectrophotometer (Shimadzu UV-2100) and the wavelength scanning range was 200~800 nm. The photocatalytic performance of the as-prepared CdS@TiO_2 heterojunction was evaluated by degradation of methyl orange solution and the exposure time was 100 min.

3. Results and discussion

3.1. XRD Analysis

Figure 1 gives the XRD patterns of TiO_2 nanobelts (a) and CdS@TiO_2 heterojunctions (b). All the diffraction peaks in the pattern of pure TiO_2 nanobelt consisted of tetragonal anatase phase and monoclinic TiO_2-B phase, which could be indexed as the JCPDS No. 21-1272 and No.35-0088, respectively. In comparison with the diffraction profile of pure TiO_2 nanobelt, a typical diffraction peak appears at 26.5° in the diffraction pattern of CdS@TiO_2 heterojunction, which can be indexed as the hexagonal CdS and its JCPDS files was nos. 80-0006. The broad nature of the CdS XRD peaks indicates that the sizes of the CdS nanoparticles are very small.

3.2. Morphology Analysis

Figure 2 exhibits the typical TEM(a), HRTEM(b) and SEM(c-d) images of CdS@TiO_2 heterojunction. As shown in Figure 2a, CdS nanoparticles are deposited and equally distributed on the surface of the nanobelts. From the HRTEM image(b), it can be seen that the size of CdS nanoparticles is about 10 nm. Figure 2c is the typical SEM image, which also illustrates CdS nanoparticles grow on the surface of the TiO_2 nanobelts. The SEM image of a single heterostructured nanobelt is shown in Figure 2d. It can be seen that CdS particles with grain size about 10 nm are also deposited on all the sides of the nanobelt. The SEM results is in good agreement with the TEM observation. From the cross section of the nanobelt, the width of the nanobelt is 250 nm and the thickness is about 50 nm.

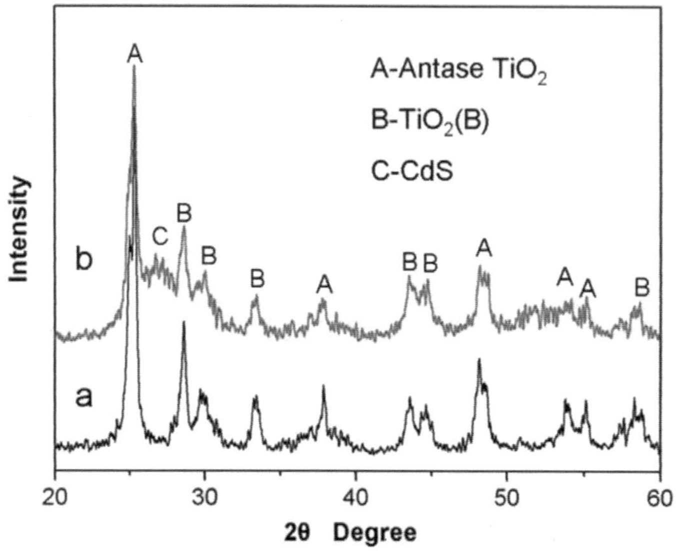

Figure 1. XRD patterns of TiO_2 nanobelts (a) and CdS@TiO_2 heterojunction (b).

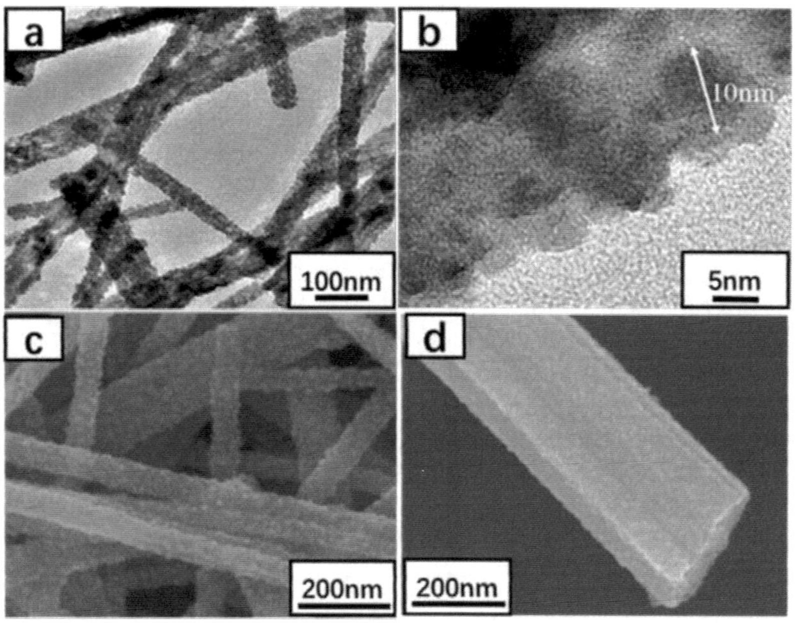

Figure 2. The typical TEM(a), HRTEM(b) and SEM images(c-d) of CdS@TiO_2 heterojunction.

3.3. UV-vis Analysis

The UV-vis spectra of TiO_2 nanobelts and CdS@TiO_2 heterojunction were exhibited in Figure 3. The pure TiO_2 nanobelts exhibit a steep absorption edge located at about 390 nm. The CdS@TiO_2 heterojunction displays strong capability of light absorption in both UV and visible light range of 200-530 nm, which will lead to good visible light photocatalytic activity. The visible light absorption of CdS@TiO_2 heterojunction is attributed to the presence of CdS nanoparticles on the surface of TiO_2 nanobelts.

3.4 Analysis of photocatalytic performance and mechanism.

Figure 4 shows the photocatalytic degradation to methyl orange of TiO_2 nanobelts and CdS@TiO_2 heterojunction at different times (20 min, 40 min, 60 min, 80 min and 100 min). It can be seen from

Figure 4 that the degradation rate of methyl orange increase with the extension of the time for both samples. The photocatalytic effect of CdS@TiO_2 heterojunction is obviously better than that of TiO_2 nanobelt. Under UV irradiation for 100 min, the degradation rate of CdS@TiO_2 heterojunction on methyl orange reaches 64.9%, while the degradation rate of TiO_2 on methyl orange is only 35%. The reason for the improvement of photocatalytic performance is that the TiO_2 and CdS with different band gaps achieve the energy level overlap and form heterogeneous junction, which is beneficial to the transport and separation of photogenerated carriers.

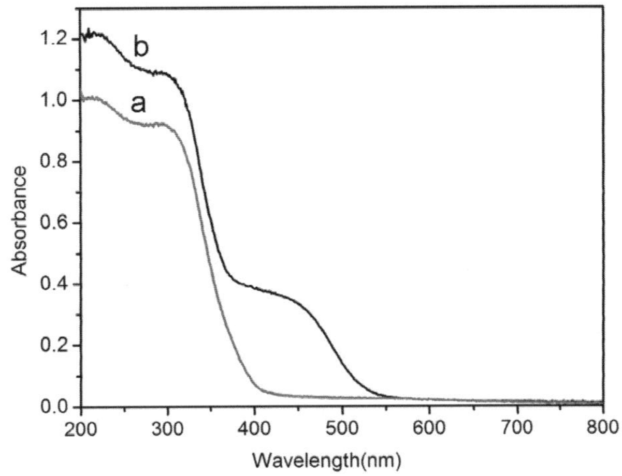

Figure 3. UV-Vis spectra of TiO_2 nanobelts and CdS@TiO_2 heterojunction.

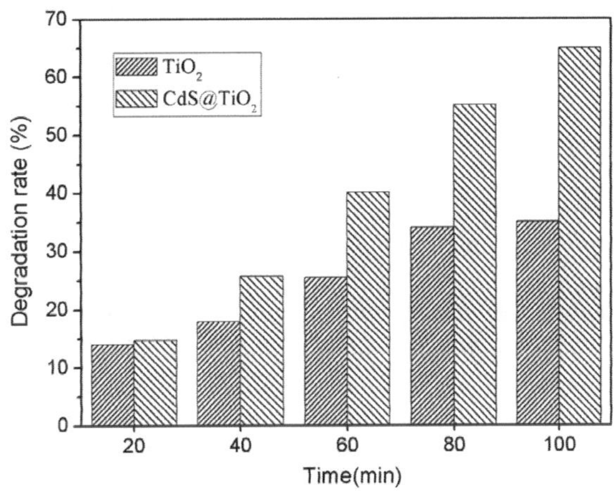

Figure 4. Photocatalytic degradation to methyl-orange solution of TiO_2 nanobelts and CdS@TiO_2 nanobelts at different times.

Figure 5 shows schematic diagram of the photogenerated carriers transport and separation path. When the TiO_2 and CdS are exposed to ultraviolet light or excited with photons of sufficient energy, the electrons on TiO_2 and CdS valence-bands simultaneously are generated and jumped to the corresponding conduction band. Due to the different energy levels in the conduction band and valence band between the two semiconductors, photogenerated electrons are concentrated in the conduction band of TiO_2, while the holes are concentrated in the CdS valence band. As thus, the photogenerated carriers are separated and the quantum efficiency can be improved.

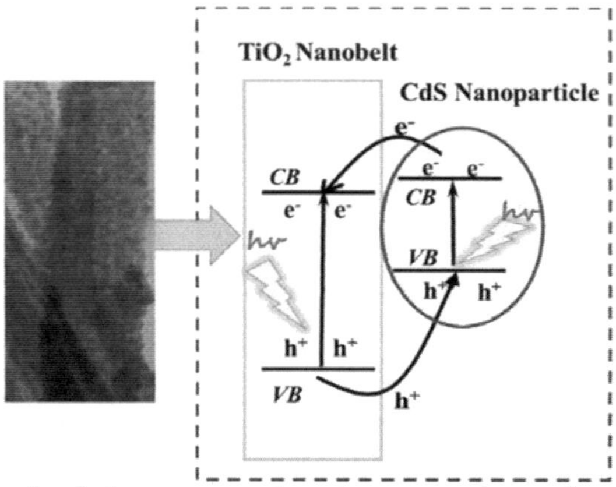

Figure 5. The schematic of photogenerated carriers separation and transport path in CdS@TiO_2 nanobelts.

4. Conclusions

In this study, TiO_2 nanobelts, prepared by hydrothermal method and subsequent heat treatment method, was used as substrate, and CdS nanoparticles were deposited on the surface in situ by liquid phase synthesis method. In this method, the CdS@TiO_2 nanoparticles-nanoribbon Heterojunction was obtained. The photocatalytic efficiency of CdS@TiO_2 was much higher than that of pure TiO_2 nanobelt. When the degradation time of methyl-orange solution was 100 min, the degradation rate of CdS@TiO_2 was 1.9 times as high as that of pure TiO_2 nanobelt, which was attribute to the realizing of band gap interleaving of the CdS and TiO_2 semiconductors. Thereby the photogenerated carriers (electrons and holes) are effectively separated and the quantum efficiency can be improved.

5. References

[1] Hoffmann M R, Martin S T, Choi W and Bahnemann D W 1995 *Chemical reviews* **95** 69
[2] Wang H, Zhang L, Chen Z, Hu J, Li S, Wang Z, Liu J and Wang X 2014 *Chemical Society Reviews* **43** 5234
[3] Tada H, Mitsui T, Kiyonaga T, Akita T, Tanaka K 2006 *Nature materials* **5** 782.
[4] Wang S, Gao Y, Miao S, Liu T, Mu L, Li R, Fan F and Li C 2017 *Journal of the American Chemical Society* **139** 11771
[5] Zhang M, Chen C, Ma W and Zhao J 2008 *Angewandte Chemie International Edition* **47** 9730
[6] Lee J C, Kim T G, Lee W, Han S H and Sung Y M 2009 *Crystal Growth & Design* **9** 4519
[7] Baker D R and Kamat P V 2009 *Advanced Functional Materials* **19** 805
[8] Banerjee S, Mohapatra S K, Das P P and Misra M 2008 *Chemistry of materials* **20** 6784
[9] Wang G L, Xu J J, Chen H Y and Fu S Z 2009 *Biosensors and Bioelectronics* **25** 791
[10] Xian J, Li D, Chen J, Li X, He M, Shao Y, Yu L and Fang J 2014 *ACS applied materials & interfaces* **6** 13157
[11] Kuai L, Zhou Y, Tu W, Li P, Li H, Xu Q, Tang L, Wang X, Xiao M and Zou Z 2015 *RSC Advances* **5** 88409
[12] Lian Z, Xu P, Wang W, Zhang D, Xiao S, Li X and Li G 2015 *ACS applied materials & interfaces* **7** 4533
[13] Wang Y, Du G, Liu H, Liu D, Qin S, Wang N, Hu C, Tao X, Jiao J, Wang J, Wang Z. 2008 *Advanced Functional Materials* **18** 1131
[14] Mugdur P H, Chang Y J, Su Y W, Mugdur P H, Chang Y J, Su Y W, Morrone A A, Ryu S O, Lee T J and Chang C H 2007 *Journal of the Electrochemical Society* **154** 482

Acknowledgments

The research is supported by Natural Science Foundation of Shandong Province (Grant No. ZR2018BEM019), Natural Science Foundation of Jiangsu Province (BK20170221), National Natural

Science Foundation of China (Grant No.51802117), Youth science foundation of Shandong Province (Grant No. BS2011CL028) and the Shandong Jiaotong University "Climbing" Research Innovation Team Program.

IOP Conf. Series: Materials Science and Engineering **617** (2019) 012004 doi:10.1088/1757-899X/617/1/012004

Hierarchical Urchin-like Spinel $Cu_xCo_{3-x}O_4$ Particles as Oxygen Evolution Reaction Catalysts in Alkaline Medium

R A E Acedera, and M D L Balela

Sustainable Electronic Materials Group, Department of Mining, Metallurgical, and Materials Engineering, University of the Philippines, Diliman, Quezon City, Philippines 1101

E-mail: rose_anne.acedera@upd.edu.ph, mlbalela1@up.edu.ph

Abstract. In this work, spinel $Cu_xCo_{3-x}O_4$ (x = 0.00, 0.50, 1.00, 1.50) was synthesized through a typical ethanol-assisted hydrothermal route. The effect of the incorporation of copper on the resulting morphology, structure, and oxygen evolution reaction (OER) activity of the spinel Co oxides was investigated. In general, Cu-doped samples exhibited improved OER activity associated to the reduction of electrical resistivity of the spinel due to the presence of Cu^{2+} into the lattice. Further, the ionic bonding of OH^- with Cu^{2+} is weaker relative to the Co^{3+} ions at the octahedral sites, which could have facilitated OER better. Among the Cu concentration experimented, $Cu_{0.5}Co_{2.5}O_4$ exhibited the best OER catalytic with an onset overpotential and Tafel slope of about 271 mV and 100.4 mV·dec^{-1}, respectively. This is attributed to its better conductivity due to the absence of a resistive secondary CuO phase.

1. Introduction

Oxygen evolution reaction (OER) has been a central process in most energy storage and conversion systems today [1]. However, regardless of the nature of the medium used, it proceeds through a four-step electron transfer mechanism. Its complex nature results to slow reaction kinetics and high electrochemical overpotentials, thereby drastically lowering the efficiency of energy conversion [2, 3]. To resolve such drawbacks, catalysts are used in practice to facilitate OER. At present, catalysts based on precious metals such as Pt [4], Ir [5], and Ru [6] exhibited superior electrocatalytic activities. However, their high cost, scarcity, and limited stability in alkaline environment hinder their industrial application [7, 8].

Alternative OER electrocatalysts include those which are based on earth abundant transition metals, such as spinel oxides [9], perovskites [10], layered double hydroxides [11], chalcogenides [12], carbides [13], and sulfides [14]. Particularly, spinel transition metal oxides showed promising electrocatalytic activity due to their good electrical conductivities, and the presence of active sites for oxygen adsorption and release in their structures. These resulted from the efficient electron transfer between the metals of different valences in the spinel structure [15, 16]. Among these spinel transition metal oxides, cobalt (II, III) oxide (Co_3O_4) is regarded as one of the best heterogeneous OER electrocatalysts with its excellent activity, and stability, even in alkaline electrolytes [17-20]. It is the most stable oxide of cobalt, consisting of Co^{2+} and Co^{3+} ions in the tetrahedral and octahedral sites, respectively.

To further improve its intrinsic OER activity, Co_3O_4 has been doped with different metals. For example, Co_3O_4 was doped with Li, which resulted in lower electrical resistivity and onset potential by about 50 mV [21]. On the other hand, Ni-doped Co_3O_4 showed better OER activity than the undoped oxide even at Ni-doping as low as 0.25 atomic % [22]. The increased activity was observed up to 0.50% Ni-doping ($Co_{2.985}Ni_{0.015}O_4$). However, it was reduced when the amount of Ni was further increased [22]. Moreover, a comparative study on the catalytic behavior of $M_xCo_{3-x}O_4$ (M = Li, Ni, Cu) showed that OER activity increased in the order $Co_3O_4 < Ni_xCo_{3-x}O_4 << Cu_xCo_{3-x}O_4 < Li_xCo_{3-x}O_4$, indicating that Cu is a viable dopant over other metals due to its cost-effectivity [23].

A number of studies has reported OER activity enhancement of Co_3O_4 through Cu-doping. However, most of these articles made use of $Cu_xCo_{3-x}O_4$ thin films or coatings [24-25]. Specifically, the use of hierarchically-structured $Cu_xCo_{3-x}O_4$ particles as OER electrocatalyst has not been reported elsewhere. In the present paper, hierarchical urchin-like, spinel $Cu_xCo_{3-x}O_4$ nanostructures were synthesized using ethanol-assisted hydrothermal route and tested for OER catalysis. The effects of copper incorporation on the morphology, structure and catalytic activity of Co_3O_4 were investigated.

2. Experimental details

2.1. Catalyst synthesis

The spinel $Cu_xCo_{3-x}O_4$ electrocatalysts were synthesized through a typical ethanol-assisted hydrothermal route as reported previously [26]. All of the chemical reagents were analytical grade and used without further purification. In a typical procedure, cobalt nitrate hexahydrate [$Co(NO_3)_2 \cdot 6\ H_2O$, Sigma-Aldrich, 98%] and copper nitrate hemi pentahydrate [$Cu(NO_3)_2 \cdot 2.5\ H_2O$, Sigma-Aldrich, 98%] were mixed accordingly to prepare four compositions of $Cu_xCo_{3-x}O_4$ (x = 0.00, 0.50, 1.00, and 1.50). The total amount of metal salts was held constant at 5.00 mmol. The metal salts were added into a solution of 1.00 g cetyltrimethylammonium bromide (CTAB, $C_{19}H_{42}BrN$, Loba Chemie, 99%) and 2.00 g urea (CH_4N_2O, Techno Pharmchem, 99.5%) in a 1:2 ethanol-water mixed solvent [10.00 mL anhydrous ethanol (C_2H_5OH, Ajax Finechem, 100%) and 20.00 mL distilled water]. To ensure homogeneous mixing, the solutions were stirred at 300 rpm for 1 h, and then transferred into 100 mL Teflon-lined steel autoclaves. These were held at 100 °C for 10 h. After the reaction, the precipitates were collected via centrifugation, washed three times with deionized H_2O and ethanol, and then dried at 100 °C for 2 h. Calcination was done at 400 °C for 6 h, at a heating rate of 1 $C^o \cdot min^{-1}$.

2.2. Material characterization and electrochemical studies

The resulting morphologies of the samples was examined in a field-emission scanning microscope (FESEM, Hitachi SU 8230). Crystal structure and phase identification were performed by X-ray diffraction (XRD, PanAlytical X'pert PRO MPD, Cu $K\alpha$).

Electrochemical measurements were done on a Biologic S-150 instrument using a three-electrode cell setup, with $Cu_xCo_{3-x}O_4$/ carbon fiber paper (CFP), Ag/AgCl (MetrOhm), and Pt wire as the working, reference, and counter electrodes, respectively. The working electrode was prepared by depositing the catalyst ink, made up of a dispersion of 80.00 mg active material, and 20.00 mg polytetrafluoroethylene (PTFE, $(C_2F_4)_n$, Sigma-Aldrich) binder in minute amount of anhydrous ethanol, onto previously cleaned 1.50 x 1.00 cm^2 CFP. The mass loading of the active material was ~ 0.50 $mg \cdot cm^{-2}$.

All electrochemical experiments were done at room temperature, ambient pressure, and in 1.00 M KOH electrolyte (pH = 13.81). OER polarization curves were obtained through linear sweep voltammetry (LSV) from 0.10 to 1.10 V (vs. *Ag/AgCl* electrode) at a scan rate of 1.00 $mV \cdot s^{-1}$. On the other hand, the stability of the working electrode was assessed by subjecting it to 500 cycles of polarization from 0.40 to 0.70 V (vs. *Ag/AgCl*) at a rate of 100 $mV \cdot s^{-1}$. To further ascertain the stability of the electrodes, chronoamperometry (CA) and chronopotentiometry (CP) were done by applying 0.60 V (vs. *Ag/AgCl*), and 1.00 $mA \cdot cm^{-2}$ current density onto the working electrode for 2 h.

All obtained potential values were converted against the reversible hydrogen electrode (*RHE*) using the equation $E_{RHE} = E_{Ag/AgCl} + 0.0592\ pH + E^{o}_{Ag/AgCl}$ where E_{RHE} is the converted potential with respect to *RHE*, $E_{Ag/AgCl}$ is the experimental potential value obtained, and $E^{o}_{Ag/AgCl}$ is the standard reduction potential of the *Ag/AgCl* (3.30 M KCl) electrode at 25 °C. Subsequently, the overpotential values were then calculated with respect to the thermodynamic potential equal to 1.23 V (vs. *RHE*).

3. Results and Discussion

3.1. Material characterization

Figure 1a shows the XRD patterns of the synthesized $Cu_xCo_{3-x}O_4$ powders. For the undoped sample (x = 0.00), the peaks identified at 2θ = 18.99, 31.33, 36.86, 38.72, 44.89, 55.67, 59.34, 65.26, and 77.4° can be well-indexed to the (111), (220), (311), (222), (400), (422), (511), (440), and (531) planes of spinel Co_3O_4 (JCPDS 43-1003) [27]. On the other hand, the diffraction pattern of $Cu_{0.5}Co_{2.5}O_4$ (x = 0.50) is identical to that of the parental spinel Co_3O_4, showing no additional peaks associated to secondary phases. This indicates that all Cu^{2+} ions have been incorporated entirely into the oxide, and their substitution with Co^{2+} did not alter the spinel structure [22]. However, as the amount of Cu was increased (x = 1.00, and 1.50), additional peaks at 2θ = 35.58, 48.71, and 61.5° appeared. These peaks were attributed to the (002), (202), and (113) planes of monoclinic CuO, which formed as a secondary phase [28]. In fact, the amount of CuO apparently increased as x was changed from 1.00 to 1.50. As seen in Figure 1b, the peak at around 38.50°, which are due to both (222) spinel oxide and (111) CuO peaks increased in intensity. These are in agreement with previous reports that pure Cu-Co oxide can be obtained at x values less than 1.00 [25].

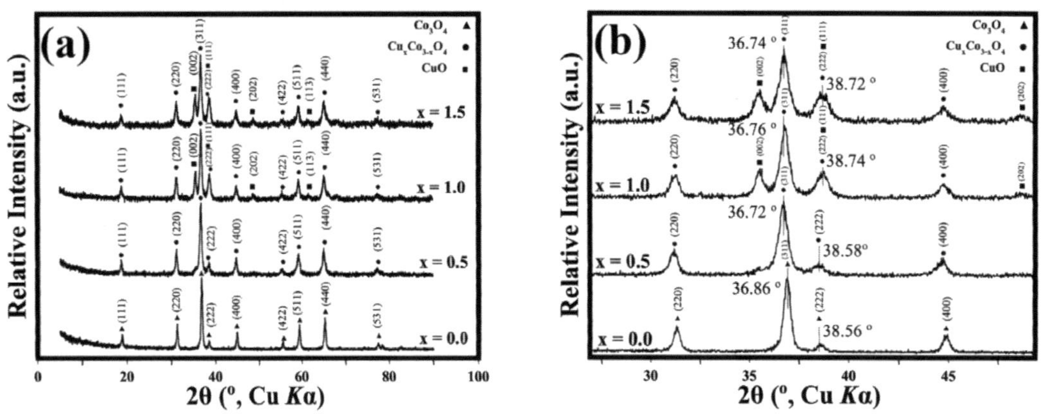

Figure 1. (a) XRD patterns of $Cu_xCo_{3-x}O_4$ and, (b) magnified view of the plot from $2\theta = 30^{o}$ to 47.5°

Table 1 lists down the structural parameters calculated for the oxides. For all samples, peak broadening was observed, indicative of their small crystallite size. The crystallite size was calculated using the Debye-Scherrer approximation from the broadening of the (311) peak. On the other hand, it is clear from Figure 1b that peak positions shifted to lower 2θ values as Cu was incorporated. These small displacements imply enlargement of the crystal lattice, reflected by the increasing lattice parameter values in Table 1. This could be ascribed to the larger radii of Cu^{2+} (0.73, 0.57 Å) in contrast to that of Co^{2+} and Co^{3+} (0.52, 0.58 Å) in the octahedral, and tetrahedral sites of the spinel [22, 29-30]. On the other hand, the calculated lattice parameter of the undoped and Cu-doped sample is comparable with the reported values for spinel Co_3O_4 (8.08 Å) [31], and of $Cu_{0.92}Co_{2.08}O_4$ (8.10 Å), respectively [25].

Table 1. Structural parameters of $Cu_xCo_{3-x}O_4$

Sample	2θ [°]	*d* spacing [Å]	Lattice Parameter, *a* [nm]	FWHM [°]	Crystallite size [nm]
x = 0.00	36.860	2.436	8.078	0.380	22.036
x = 0.50	36.720	2.445	8.108	0.520	16.097
x = 1.00	36.760	2.442	8.099	0.580	14.433
x = 1.50	36.740	2.444	8.104	0.560	14.948

Figure 2 shows the FESEM images of the synthesized $Cu_xCo_{3-x}O_4$ oxides. The undoped Co_3O_4 are spherical with a mean diameter of 63.507 nm (SD = 14.626) as seen in Figure 2a-b. The particles, however, agglomerated to reduce to their high surface energy due to their small particle size. Upon Cu-incorporation, a significant change in morphology was observed. Urchin-like microparticles, composed of nanowires of increasing mean diameter of 13.351 (SD = 66.664), 26.662 (SD = 32.273), and 31.482 nm (SD = 20.643), were produced for x = 0.50, 1.00, and 1.50, respectively. The formation of urchin-like particles can be explained by the anisotropic growth promoted by the incorporation of the dopant Cu^{2+} ions. Anisotropic growth results from the difference in the adsorption energies of the dopant on the different facets of the host spinel Co_3O_4 [32].

The formation of nanosheets can also be observed at x = 1.00 and became more evident at x = 1.50. These nanosheets could be the secondary CuO phase detected by XRD at excess Cu. The CuO nanosheets were possibly formed from $Cu(OH)_2$, which were generated from the hydrolysis of Cu^{2+} ions in a strong alkaline solution. Calcination at 400 °C dehydrates $Cu(OH)_2$, thereby producing CuO. Release of water results to the formation of O-Cu-O bridges between $Cu(OH)_2$ layers. This leads to oriented attachment of the particles, thereby forming a sheet-like structure [33-34].

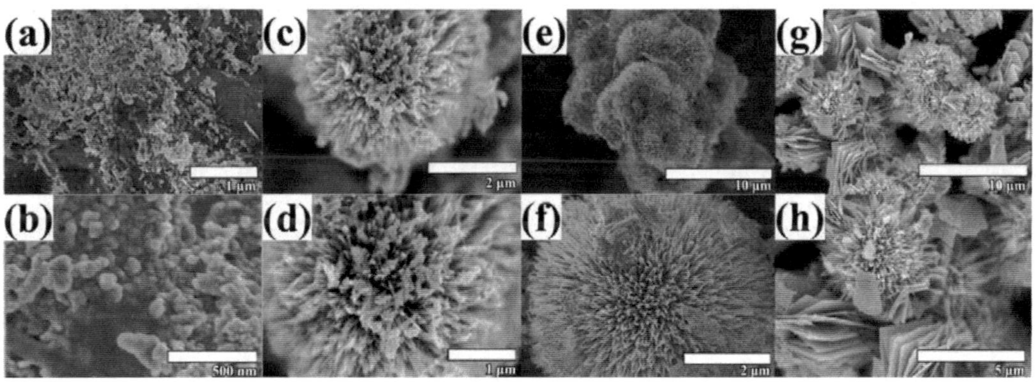

Figure 2. Low and high magnification FESEM images of $Cu_xCo_{3-x}O_4$ with (a, b) x = 0.00, (c, d) x = 0.50, (e, f) x = 1.00, and (g, h) x = 1.50

3.2. Electrochemical properties of the synthesized Co_3O_4 and $Cu_xCo_{3-x}O_4$ powders

Figure 3a shows the LSV polarization curves from 1.10 to 2.20 V (vs. *RHE*) of the Co_3O_4 and $Cu_xCo_{3-x}O_4$/CFP electrode in 1.00 M KOH. Undoped Co_3O_4 exhibited the largest onset potential for OER of 1.637 V and a corresponding overpotential of 407 mV. Upon Cu-doping, at x = 0.50, higher OER activity was achieved as the onset potential and the overpotential decreased to 1.501 V, and 271 mV, respectively. This can be attributed to the decrease of the electrical resistivity of the oxide, and the stronger bond of Co^{3+} to OH^- over Cu^{2+} [35]. M-OH bond strength was found to be a critical parameter for OER activity as it determines the ease of breaking up the bond between the metal and the OH^- ions adsorbed at the beginning of the reaction. A stronger M-OH bond needs higher energy to dissolve the bond, leading to larger overpotential. On the other hand, as the amount of Cu was further increased, the onset potential shifted more positively to 1.531 (x = 1.00) and 1.542 V (x =1.50). Although the catalytic activity improved compared to the undoped Co_3O_4, the enhancement was offset due to the presence of a resistive CuO phase in the samples [36].

The abrupt increase in the measured current densities after the onset indicated the beginning of OER. The required potential to reach a 10 $mA{\cdot}cm^{-2}$-current density, a metric significant in solar fuel production, were found to be 1.520, 1.643, 1.680, and 1.696 V at x = 0.00, 0.50, 1.00, and 1.50, respectively. Correspondingly, Tafel measurements were determined from the LSV polarization data. As shown in Figure 3b, the slope of the undoped sample was 102.1 $mV{\cdot}dec^{-1}$, and those of the doped oxides were 100.4 (x = 0.50), 103.2 (x = 1.00), and 107.4 $mV{\cdot}dec^{-1}$ (x = 1.50). These values further prove the high activity of $Cu_{0.5}Co_{2.5}O_4$ over Co_3O_4. The sustained linear behavior even at high current density regions imply fast electron transfer and mass transport across the electrode-electrolyte interface [37].

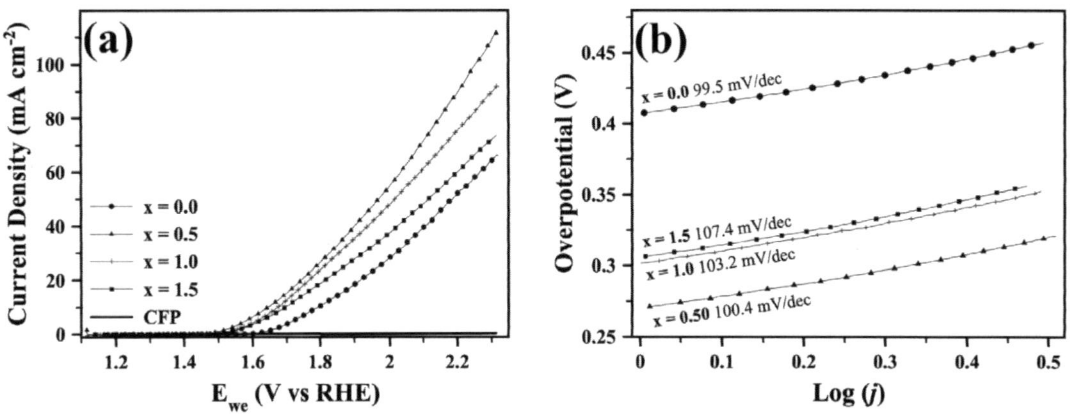

Figure 3. Electrochemical characterization of the $Cu_xCo_{3-x}O_4$/CFP electrode with (a) LSV polarization curves from 1.10 to 2.20 V vs. *RHE*, and corresponding (b) Tafel plots,

Figure 4a shows the cyclic voltammograms of the samples from 1.20 to 1.55 V (vs. *RHE*). All the electrodes exhibited characteristic redox peaks, which could be attributed to the Co^{3+}/Co^{4+} redox couple [24]:

$$CoOOH + OH^- \leftrightarrow Co\text{-}O_2 + e^- + H_2O \quad (1)$$

As the applied potential increases, energy becomes sufficient such that the bond between Co^{4+} ions and the adsorbed O_2 is severed. This bond breaking then results to the release of O_2 gas [38]. A less positive anodic peak value, therefore, indicates better OER catalytic property because it implies earlier OER onset. Herein, the anodic peak potential values at x = 0.00, 0.50, 1.00, and 1.50 are 1.506, 1.463, 1.467, and 1.470 V (vs. *RHE*), respectively. These values correspond to the Co^{3+}/Co^{4+} redox couple peak positions for $Cu_xCo_{3-x}O_4$ electrodes (1.43 to 1.53 V vs *RHE*) [24, 39-40]. Similarly, $Cu_{0.5}Co_{2.5}O_4$ exhibited the smallest anodic peak potential, which agrees with the results of the LSV. Moreover, the onset potential for all electrodes shifted in correspondence to the variation of the position of the Co^{3+}/Co^{4+} redox peak [24].

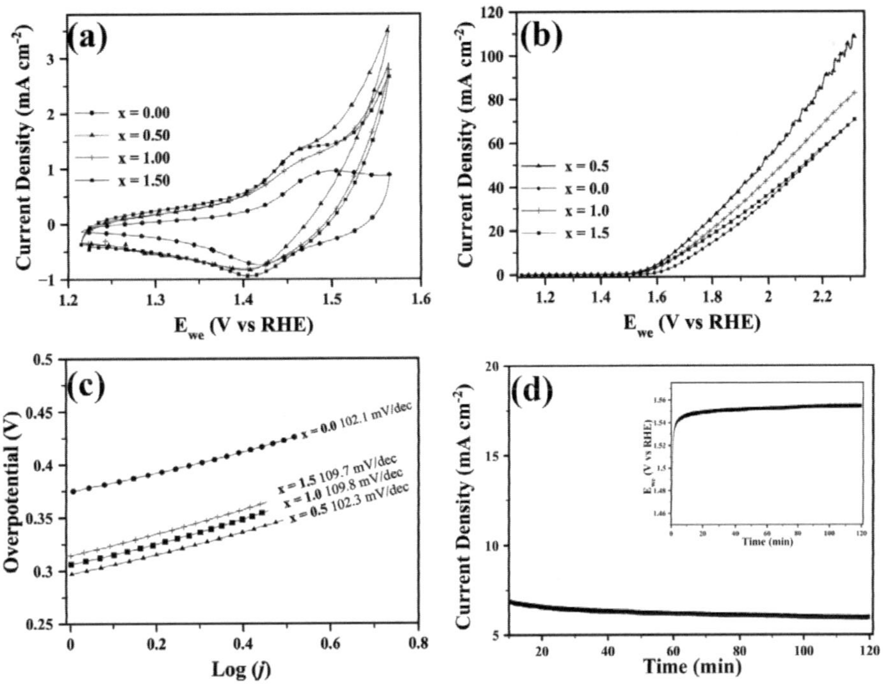

Figure 4. (a) Cyclic voltammograms from 1.20 to 1.55 V vs. *RHE*, (b) LSV plots after 500 cycles of polarization, and corresponding (c) Tafel plots, and (d) CA plot of the $Cu_{0.5}Co_{2.5}O_4$ electrode (inset: CP plot)

To determine the stability of the OER electrodes, 500 cycles of polarization in the potential window 1.40 to 1.80 V were done. Figure 4b-c show the LSV curves and the corresponding Tafel plots of the samples before and after the stability test. The onset potentials increased by 2.70 to 2.56% at x = 0.00 to 1.50. Correspondingly, the Tafel slopes were also larger by 2.45 to 2.10%. These prove that the $Cu_xCo_{3-x}O_4$/CFP electrodes remained stable even after numerous cycles of polarization. Furthermore, it can be seen in Figure 4d that the measured chronoamperometric (~ 6 $mA \cdot cm^{-2}$), and chronopotentiometric response (1.55 V) of $Cu_{0.5}Co_{2.5}O_4$/CFP remained constant even after 2 h, establishing further the stability of the best performing electrode.

4. Conclusion

In conclusion, spinel $Cu_xCo_{3-x}O_4$ (x = 0.00, 0.50, 1.00, and 1.50) powders were synthesized through ethanol-assisted hydrothermal route. Incorporating Cu^{2+} ions onto the spinel Co_3O_4 lattice resulted to a drastic change in morphology from agglomerated, spherical particles to micro urchins composed of nanowires. Furthermore, doping beyond x = 0.50 yielded a secondary CuO phase. For the OER catalytic activity testing, Cu^{2+} doping caused improvements to the performance of Co_3O_4. $Cu_{0.5}Co_{2.5}O_4$ was found to be the best performing electrode with onset overpotential and Tafel slope as low as 1.501 V, and 100.4 $mV \cdot dec^{-1}$, respectively. Increasing Cu content beyond x = 0.50 resulted to lower OER activity due to the presence of a resistive CuO phase. Stability tests proved the durability of the $Cu_xCo_{3-x}O_4$/CFP electrodes, with the observed minimal change in their onset potentials, and Tafel slopes after 500 polarization cycles from 1.4 to 1.8 V.

Acknowledgment

This study is supported by the Commission on Higher Education (CHED)-Newton Agham Institutional Links, under the project *"Affordable Electrolyzer Technology based on Transition Metal*

Catalyst for Energy Storage" and Robert Cheng – Uratex 2019 College of Engineering Professorial Chair. Funding for conference attendance was also provided by UP College of Engineering.

References

[1] Doyle S R L and Lyons M E G The Oxygen Evolution Reaction: Mechanistic Concepts and Catalyst Design 2010 (Switzerland)
[2] Hu A T, Wang Y, Zhang L and Tang T 2019 *Applied Catal. B, Environ.* **243** 175–182
[3] Li R, Zhou D, Luo J, Xu W, Li J, Li S and Cheng 2017 *J. Power Sources* **341** 250–256
[4] Sung M and Kim J 2018 *Int. J. Hydrogen Energy* **43** 2130-2138
[5] Papaderakis A, Tsiplakides D, Balomenou S and Sotiropoulos S 2015 *J. Electroanal. Chem.* **757** 216–224
[6] Ma Z, Zhang Y, Liu S, Xu W, Wu L, Hsieh Y C, Liu P, Zhu Y, Sasaki K, Renner J N, Ayers K E, Adzic R R and Wang J X 2018 *J. Electroanal. Chem.* **819** 296–305
[7] Guo Y, Guo D, Ye F, Wang K and Shi Z 2017 *Int. J. Hydrogen Energy* **42** 17038–17048
[8] Si C, Zhang Y, Zhang C, Gao H, Ma W, Lv L and Zhang Z 2017 *Electrochim. Acta.* **245** 829–838
[9] Bocca C 1999 *Int. J. Hydrogen Energy* **24** 699–707
[10] Han B, Grimaud A, Giordano L, Hong W T, Diaz-Morales O, Lee Y, Hwang J, Charles N, Stoerzinger K A, Yang W, Koper M T M and Shao-horn Y 2018 *J. Phys. Chem. C* **122** 8445-8454
[11] Xie T, Min J, Liu J, Chen J, Fu D, Zhang R, Zhu K and Lei M 2018 *J. Alloys Compd.* **754** 72–77
[12] Zhang T, Zhu Y and Lee J Y 2018 *J. Mater. Chem. A*
[13] Wygant B R, Kawashima K and Mullins C B 2018 *ACS Energy Lett.*
[14] Peng L, Shah S and Wei Z 2018 *Chinese J. Catal.* **39** 1575-1593
[15] Zeng K and Zhang D 2010 *Prog. Energy Combust. Sci.* **36** 307–326
[16] Zhang Y, Zhang C, Gao H, Ma W, Lv L and Zhang Z 2017 *Electrochim. Acta.* **245** 829–838
[17] Yang X, Li H, Lu A Y, Min S, Idriss Z, Hedhili M N, Huang K W, Idriss H and Li L J 2016 *Nano Energy* **25** 42–50
[18] Asadizadeh S, Amirnasr M, Meghdadi S, Fadaei Tirani F and Schenk K 2018 *Int. J. Hydrogen Energy* **10** 4922-4931
[19] Ye K H, Zhou S A, Zhu X C, Xu C W and Shen P K 2013 *Electrochim. Acta* **90** 108–111
[20] Balela M D L, Angostura T J E, Domingo A M F and Facal C G P 2018 *Key Eng. Mater.* **775** 336–341
[21] Wu X and Scott K 2013 *Int. J. Hydrogen Energy* **8** 1–7
[22] Singhal A, Bisht A and Irusta S 2018 *J. Electroanal. Chem.* **823** 482–491
[23] Nikolov I, Darkaoui R, Zhecheva E, Stoyanova R, Dimitrov N and Vitanov T 1997 *J Electroanal Chem*
[24] Wu X and Scott K 2011 *J. Mater. Chem.* **21** 12344-12351
[25] Chi B, Lin H and Li J 2008 *Int. J. Hydrogen Energy* **33** 4763–4768
[26] Wen W, Wu J M and Tu J P 2012 *J. Alloys Compd.* **513** 592–596
[27] Zhang Y, Zhou X, Zhang F, Tian T, Ding Y and Gao H 2017 *J. Catal.* **352** 246–255
[28] Barkaoui S, Dhaouadi H, Kouass S and Touati F 2015 *Int. J. Light Electron Opt.* **126** 1047–1051
[29] Franken T and Palkovits R 2015 *Appl. Catal. B Environ.* **176** 298–305
[30] Xu Q, Su Y, Wu H, Cheng H, Guo Y, Li N and Liu Z 2014 *Curr. Nanosci.* **11** 107–112
[31] Ibupoto Z H, Khun K, Beni V, Liu X and Willander M 2013 *SNL* 7926–7938
[32] Han X, Wahl S, Russo P and Pinna N 2018 *Nanomaterials* **8** 249
[33] Cudennec Y, Lecerf A 2003 *Solid State Sci* **5** 1471–1474
[34] Shahmiri M, Ibrahima N A, Zainuddin N, Azim N, Bakhtyar B, Zaharim A and Sopian K 2013 *WSEAS Transactions on Environment and Development* 243–252

[35] Wu J, Huang X, Berglund K, Lu X, Xiaohua, Feng X, Larsson R and Shi Y 2018 *CSTE* **162** 86-92
[36] De Koninck M, Poirier S and Marsan B 2006 *J. Electrochem. Soc.* A2103-2110
[37] Park Y S, Park C S, Kim C H, Do Kim Y, Parkand S and Lee J H 2016 *J. Korean Phys. Soc.* **69** 1187–1190
[38] Pittkowski R, Krtil P and Rossmeisl J 2018 *Curr. Opin. Electrochem.* **12** 218 – 224
[39] Fradette N and Marsan B 1998 *J. Electrochem. Soc.* **145** 2320– 2327
[40] La Rosa-Toro A, Berenguer R, Quijada C, Montilla F, Morallón E and Vàzquez J L 2006 *J. Phys. Chem. B* **110** 24021–24029.

IOP Conf. Series: Materials Science and Engineering **617** (2019) 012005 doi:10.1088/1757-899X/617/1/012005

Thermal Degradation of Polysiloxanes Containing Chloropropyl Groups

Fuying Dong[1], Shuaijie Li[1], Chenxing Wang[1], Meng Zhang[1] and Guoqiang Xie[1]

1 Shandong Jiaotong University, Jinan 250357, P. R. China

E-mail: dongfuying2000@aliyun.com

Abstract. We synthesized a series of polysiloxanes containing chloropropyl groups, and studied its thermal stability and degradation mechanism through thermogravimetric analysis (TGA) in nitrogen atmosphere. The effect of the introduced chloropropyl units on the thermal degradation behavior of polysiloxanes was discussed. The depolymerization of the main chain of polysiloxane was accelerated by increasing the molar content of chloropropyl groups. The introduction of chloropropyl groups can cause the formation of a cross-linking network in the residue of polysiloxanes at 800 °C and restrain the formation of volatile components.

1. Introduction

Polysiloxane possess unique characteristics, such as resistance to high temperatures and chemical stability [1, 2]. Therefore, investigating its thermal stability has important scientific and practical significance. From the point of view of science, such an investigation could gain insight into the molecular structure of polysiloxane and provide suggestions to prepare novel types of organosilicon materials. From the point of view of practicality, such an investigation can provide some inportant information for predicting the organosilicon materials used in a specific application.

Chloropropyl groups could react with many nucleophilic reagents, such as amines [3,4], hydroxy compounds [5], thiol compounds [6], carboxylates [7], cyanide [8]. Therefore, polysiloxane containing chloropropyl groups has been used as the key chemical intermediate for preparing advanced materials. However, to the best of our knowledge, no research has focused on the thermal properties of polysiloxane containing chloropropyl groups.

Thermogravimetric analysis (TGA) has been commonly utilized to investigate the thermal degradation and degradation mechanisms of materials. The aim of this work was to investigate the thermal stability and degradation mechanism of polysiloxane containing chloropropyl groups through TGA.

2. Experimental

2.1. Materials

Dimethyldiethoxylsilane, and γ-chloropropylmethyldiethoxysilane were purchased from Qufu Chenguang Chemical Co., Ltd. (China), and these materials were used as received. Octamethyltrisiloxane was obtained from Energy Chemical (China). Octamethylcyclotetrasiloxane (D_4) was utilized without further purification and was supplied by Zhejiang Xinan Chemical Industrial Group Co., Ltd. (China). Purolite CT-175 was supplied by Purolite Company (China). All other reagents were of analytical grade.

2.2. Techniques

IOP Conf. Series: Materials Science and Engineering **617** (2019) 012005 doi:10.1088/1757-899X/617/1/012005

A Bruker 400-MHz NMR spectrometer (Switzerland) was used for recording ^1H NMR spectra at room temperature with $CDCl_3$ as the solvent. TGA (TGA/DSC 1 thermal analyzer, Mettler, Switzerland) was performed in the temperature range of 30 °C to 800 °C in nitrogen atmosphere.

2.3. Preparation of polysiloxanes

2.3.1. Prepolymer of α, ω-Trimethylsiloxy-poly(methylchloropropyl)siloxane ***(I)***. A mixture of toluene (140 ml), distilled water (72 ml) and concentrated hydrochloric acid (68 ml) was placed in a glass reactor equipped with a reflux condenser, a mechanical stirrer and a thermometer. γ-Chloropropylmethyldiethoxysilane (84.8 g, 0.40 mol) was added dropwise to the solution in an ice water bath, and the reaction mixture was stirred for another 1.5 h. Then, the mixture was refluxed for 3 h. The organic layer was separated, washed with 5 wt% $NaHCO_3$ aqueous solution and distilled water, dried over $CaCl_2$, and filtered. The solvent was removed via distillation, and then 50.4 g of transparent oil was obtained. ^1H NMR (ppm, $CDCl_3$): δ = 0.02-0.32 (s, 3H, $C\boldsymbol{H}_3Si$), δ = 0.60-0.87 (t, 2H, $C\boldsymbol{H}_2CH_2CH_2Cl$), δ = 1.71-1.94 (m, 2H, $CH_2C\boldsymbol{H}_2CH_2Cl$), δ = 3.46-3.60 (t, 2H, $CH_2CH_2C\boldsymbol{H}_2Cl$).

2.3.2. α, ω-Trimethylsiloxy-poly(methylchloropropyl)siloxane ***(II)***. ***I*** (20.0 g), purolite CT-175 (0.2 g) and octamethyltrisiloxane (0.06 g, 0.25 mmol) were placed into a 100 ml three-necked, round-bottomed flask provided with a mechanical stirrer. Copolymerization of ***I*** and octamethyltrisiloxane was performed at 100 °C for 8 h to10 h. Purolite CT-175 was removed by filtration after the reaction was completed. Transparent oil (19.6 g) was obtained after the reaction mixture was devolatilized. ^1H NMR (ppm, $CDCl_3$): δ = 0.02-0.32 (s, 3H, $C\boldsymbol{H}_3Si$), δ = 0.60-0.87 (t, 2H, $C\boldsymbol{H}_2CH_2CH_2Cl$), δ = 1.71-1.94 (m, 2H, $CH_2C\boldsymbol{H}_2CH_2Cl$), δ = 3.46-3.60 (t, 2H, $CH_2CH_2C\boldsymbol{H}_2Cl$).
α, ω-Trimethylsiloxy-poly(dimethyl-methylchloropropyl)siloxanes. ***III*** (the molar content of γ-chloropropyl units is 75%) and ***IV*** (the molar content of γ-chloropropyl units is 25%), were synthesized with the similar procedure as ***II***.
The structures of ***II***, ***III*** and ***IV*** are shown in Scheme 1.

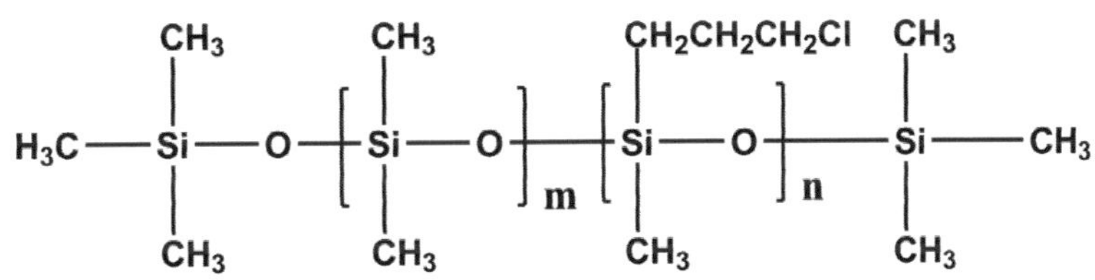

Scheme 1. Structures of ***II*** (m = 0, n ≠ 0), ***III*** (m ≠ 0, n ≠ 0, m: n = 1: 3) and ***IV*** (m ≠ 0, n ≠ 0, m: n = 3: 1).

3. Results and discussion

The thermal degradation behavior of polysiloxanes containing chloropropyl groups (***II***, ***III*** and ***IV***) were investigated through TGA in nitrogen atmosphere at heating rate of 10 °C/min.
The TG and DTG curves of ***II***, ***III*** and ***IV*** are shown in Figures 1 and 2, respectively. Figures 1 and 2 show that ***II***, ***III*** and ***IV*** exhibit many degradation shoulder peaks, indicating that the degradation mechanism of polysiloxanes containing chloropropyl groups is much complicated. Figure 2. shows that ***II*** and ***III*** decomposed rapidly at temperatures ranging from 180 °C to 400 °C in the former degradation stage, in which the peak degradation temperature of the highest rate (T_p) occurred and the largest amount of mass loss was observed. While, the degradation of polysiloxane contains small amount of chloropropyl groups in structural units (***IV***) mainly occurred in the latter degradation stage from 480 °C to 620 °C, in which T_p occurred. Thus, the introduction of large amounts of chloropropyl groups in structural units decreased the activation energy for the reaction, which implies that the

depolymerization of the main chain was accelerated by increasing the molar content of chloropropyl groups.

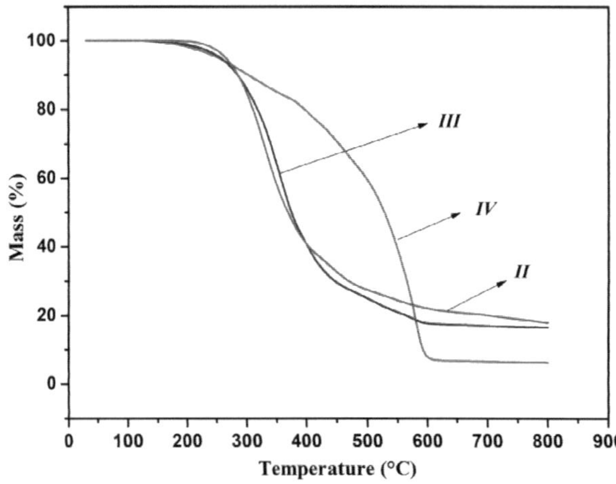

Figure 1. TG curves of ***II***, ***III*** and ***IV*** in nitrogen atmosphere.

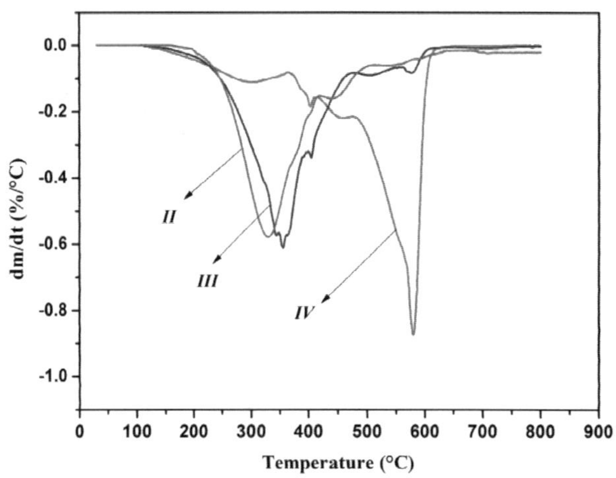

Figure 2. DTG curves of ***II***, ***III*** and ***IV*** in nitrogen atmosphere.

The characteristic degradation data of ***II***, ***III*** and ***IV*** are presented in Table 1. Table 1 shows that all the characteristic temperatures of 5% mass loss ($T_{5\ wt\%\ loss}$), 20% mass loss ($T_{20\ wt\%\ loss}$), 50% mass loss ($T_{50\ wt\%\ loss}$) and T_p of ***II***, ***III*** and ***IV*** decreased with the content of chloropropyl groups increased. This result indicated that the degradation rate of polysiloxanes containing chloropropyl groups increased with the content of chloropropyl groups increased. In addition, the residual yield of polysiloxanes containing chloropropyl groups increased with the content of chloropropyl groups in structural unit increasing at 800 °C. The residue was dipped into toluene and could not be dissolved, which indicates that the residue of polysiloxanes containing chloropropyl groups were probably cross-linked. This result indicated that the introduction of chloropropyl groups can cause the formation of a cross-linking network in polysiloxanes and restrain the formation of volatile components.

Table 1. Characteristic data of thermal degradation in nitrogen atmosphere.

Sample	$T_{5\ wt\ \%\ loss}$ (°C)	$T_{20\ wt\ \%\ loss}$ (°C)	$T_{50\ wt\ \%\ loss}$ (°C)	T_p (°C)	Residue
II	248	310	369	329	17.9
III	250	318	375	344	16.5
IV	253	398	528	580	6.3

4. Conclusion

A series of polysiloxanes containing chloropropyl groups were successfully synthesized with purolite-175 as catalyst. The effect of the introduced chloropropyl units on the thermal degradation behavior of polysiloxanes was discussed through thermogravimetric analysis (TGA) in nitrogen atmosphere. The introduction of chloropropyl groups exerted a complicated effect on the thermal degradation behavior of polysiloxanes, probably because of the synergistic effects between the thermal degradation reactions and the formation of a cross-linking network.

5. References

[1] Druvari D, Koromilas N, Lainioti G, Bokias G, Vasilopoulos G, Vantarakis A, Baras I, Dourala N and Kallitsis J 2016 Polymeric Quaternary Ammonium-Containing Coatings with Potential Dual Contact-Based and Release-Based Antimicrobial Activity ACS *Appl. Mater. Interfaces.* **8** pp 35593−605

[2] Chen L and Fan L 2018 Dendrite-free Li metal deposition in all-solid-state lithium sulfur batteries with polymer-in-salt polysiloxane electrolyte *Energy Storage Mater.* **15** pp 37–45

[3] Kobayashi M, Chiba T, Tsuda K and Takeishi M 2005 Anionic Polymerization of N, N-Dialkylacrylamides Containing Alkoxysilyl Groups in the Presence of Lewis Acids *J Polym. Sci. Pol. Chem.* **43** 2754–64

[4] Hempenius M, Robins N, Lammertink R and Vancso G 2001 Organometallic Polyelectrolytes: Synthesis, Characterization and Layer-By-Layer Deposition of Cationic Poly(ferrocenyl(3-ammoniumpropyl)-methylsilane) *Macromol. Rapid Commun.* **22** pp 30–33

[5] Dimitrov I, Jankova K and Hvilsted S 2010 Synthesis of Polystyrene-Based Random Copolymers with Balanced Number of Basic or Acidic Functional Groups *J Polym. Sci. Pol. Chem.* **48** pp 2044–52

[6] Wan D, Fu Q and Huang J 2006 Synthesis of Amphiphilic Hyperbranched Polyglycerol Polymers and Their Application as Template for Size Control of Gold Nanoparticles *J Appl. Polym. Sci.* **101** pp 509−14

[7] Feng L, Zhou L and Feng S 2016 Preparation and characterization of silicone rubber cured via catalyst-free aza-Michael reaction *RSC Adv.* **6** pp 111648–56

[8] Boullanger A, Gracy G, Bibent N, Devautour-Vinot S, Clément S and Mehdi A 2012 From an Octakis(3-cyanopropyl)silsesquioxane Building Block to a Highly COOH-Functionalized Hybrid Organic–Inorganic Material *Eur. J. Inorg. Chem.* **1** pp 143–50

Acknowledgements

This work was financially supported by PhD Research Startup Foundation of Shandong Jiaotong University, the Science Foundation of Shandong Jiaotong University (No: Z201712), the Shandong Jiaotong University "Climbing" Research Innovation Team Program, and the National Students' innovation and entrepreneurship training program (No: 201811510036).

Effects of types of isocyanate and polyol on thermal and mechanical properties of benzoxazine-urethane polymer alloys

Pichaya Akahart[1], Panuwat Luengrojanakul[2] and Sarawut Rimdusit[1]

1,2 Department of Chemical Engineering, Faculty of Engineering, Chulalongkorn University, Bangkok 10330, Thailand
E-mail: sarawut.r@chula.ac.th

Abstract. Polymer alloys of benzoxazine resin (BA-a) and urethane prepolymer (PU) were prepared at various BA-a : PU mass ratios (100 : 0, 80 : 20, 70 : 30 and 60 : 40) via sequential cure method. Isocyanate end capped urethane prepolymer was synthesized with isocyanates and polyols. Various types of isocyanate groups were used [i.e., TDI, MDI, IPDI and HDI] in this study. Polypropylene glycol and Poly(ethylene, diethylene adipate) glycol represent polyether polyol (E) and polyester polyol (S), respectively. Experimental results revealed that glass transition temperature (T_g), flexural strength and flexural modulus increase with decreasing PU content. Furthermore, the flexural strength and flexural modulus synergism were found in the BA-a:PU based polyester polyol at all mass ratio.

1. Introduction

Polyurethanes (PU) is a synthetic polymeric material used to provide several properties for versatile application, which are the results of two-phase structure separation morphology between hard domain derived from isocyanate and soft domain from long-chain diol or triol [1].

Polyurethane have excellent properties such as solvent resistance, superior toughness, and impact resistance, high tear strength and elongation, etc. However, it is not without drawback which is poor thermal stability at high temperature, limiting their application in many fields [9, 12]. Enhancement of thermal properties can be achieved by several approaches. One convenient method for effective thermal stability enhancement is to blend polyurethane with better thermal stability polymer [2].

Polybenzoxazine (PBA) is a novel class of phenolic resins which exhibits many unique properties such as high thermal stability, excellent mechanical properties, low moisture absorption, low melt viscosity, high char yield, low coefficient of thermal expansion and near-zero shrinkage Moreover, it can be synthesized via either solvent or solventless technology. The benzoxazine monomer requires no catalyst for the ring opening polymerization [2-7]. Although Polybenzoxazine (PBA) offer many excellent properties as mentioned above, its high rigidity limits their widespread use [2, 4, 6]. Hence, researchers make great effort to enhance its toughness. Another outstanding property of benzoxazine which has attracted many researchers' attention is its capability to mix with various resins by alloying or blending.

The existing phenolic hydroxyl groups in polybenzoxazine were considered to be a good match for polyurethane by reacting with the NCO groups in PU prepolymer, forming cross-linked copolymers. Therefore, considering the aforementioned statements, alloying urethane/benzoxazine is considered as the most promising approach. Not only does this display improvement of thermal stability in polyurethane but also the toughening of polybenzoxazine led them to be used as high performance polymer alloy in a wide range of applications [2, 3, 5].

2. Experimental

2.1. Material.

Sources of the Materials used for the synthesis of benzoxazine monomer and urethane pre-polymer are listed in Table 1.

Table 1. Materials used for the synthesis of benzoxazine monomer and urethane pre-polymer.

	Chemical	Source
BA-a resin	Bisphenol A *(polycarbonate grade)*	PTT Phenol Co., Ltd (Thailand).
	Formaldehyde *(AR grade)*	Merck Company
	Aniline	Loba Chemie Pvt. Ltd.
PU prepolymer	Polypropylene glycol *(Mn 1000 g/mol)*	IRPC Polyol Co., Ltd.
	Poly(ethylene, diethylene adipate) glycol *(Mn 1000 g/mol)*	Foamtec International Co., Ltd.
	2,4-toluene diisocyanate (TDI)	Foamtec International Co., Ltd.
	Methylene diphenyl diisocyanate (MDI)	Foamtec International Co., Ltd.
	Isophorone diisocyanate (IPDI)	Vencorex (Thailand) Co., Ltd.
	Hexamethylene diisocyanate (HDI)	Vencorex (Thailand) Co., Ltd.

2.2. Specimen preparation

2.2.1. Isocyanate end capped urethane prepolymer synthesis. The urethane prepolymer was synthesized via mixing excess isocyanate with polyols at a molar ratio of 1.5 : 1, and varying the type of isocyanate and polyol components used. Components and conditions used in the synthesis of PU prepolymer are given in Table 2.

Table 2. Component and condition synthesis for PU prepolymer.

Urethane prepolymer	Condition	
	Time (h)	Temperature (°C)
PU-TDI/Polyol	1.30	70
PU-MDI/Polyol	1	45
PU-IPDI/Polyol	2	90
PU-HDI/Polyol	2	90

After the reaction completed, the obtained urethane prepolymer was purged with nitrogen gas before closing the lid. It was then left at room temperature to cool down then kept in a refrigerator for future use.

2.2.2. Benzoxazine monomer (BA-a) synthesis. BA-a monomer was synthesized by using bisphenol-A, formaldehyde and aniline at a molar ratio of 1 : 4 : 2 by solventless technology. The resin was a transparent yellow solid at room temperature. It was grounded into powder before alloying with PU prepolymer.

2.2.3. Benzoxazine-urethane polymer alloys preparation. Polymer alloy comprises of a mixture of BA-a and urethane prepolymer at various mass fraction (100 : 0, 80 : 20, 70 : 30 and 60 : 40). Both of the components were heated at 125 °C in an aluminum pan and stirred to yield a homogeneous mixture. The molten resin mixture was step-cured in an air-circulated oven. After the curing process, the samples were cut into various shapes for each specific characterization.

3. Characterization

3.1. Dynamic mechanical analysis (model DMA242, Netzsch, Germany)

was carried out to determine the thermo-mechanical behavior of the specimen dimension of 10 mm×50 mm×1.5mm by using three-point bending mode. The testing was evaluated at a heating rate of 2 °C min^{-1} from 30 °C–250 °C under nitrogen atmosphere.

3.2. *Flexural Property Measurement.*

The flexural properties of the BA-a:PU alloy specimens were determined using a universal testing machine (model 5567) from Instron Co., Ltd. Following ASTM D 790M-93 (three point bending mode) with a support span of 32 mm using a constant crosshead speed of 1 mm/min. The sample dimension was 10 mm×50 mm×1.5 mm.

4. Results and discussion

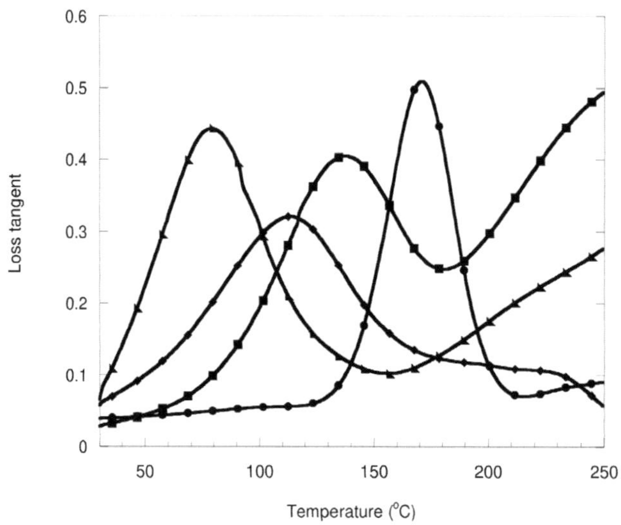

Figure 1. Loss tangent of BA-a : PU-TDI/S polymer alloys at various mass ratio: 100 : 0 (●), 80 : 20 (■), 70 : 30 (◆) and 60 : 40 (◣).

Table 3. Glass transition temperatures of neat BA-a and BA-a : PU-TDI/S polymer alloys.

BA-a:PU-TDI/S mass ratio	T_g (°C)	T_g range (°C)	ΔT_g (°C)
100:10	170	130 to 190	60
80:20	135	70 to 170	100
70:30	120	40 to 180	140
60:40	80	30 to 140	110

Table 3 presents interpreted data of tan δ curve per Figure 1. The result shows that the range of T_g (ΔT_g) increases as urethane content in polymer alloy increases. Previous researches suggest the appropriate range of T_g should be about 75 °C– 110 °C for shape memory application.

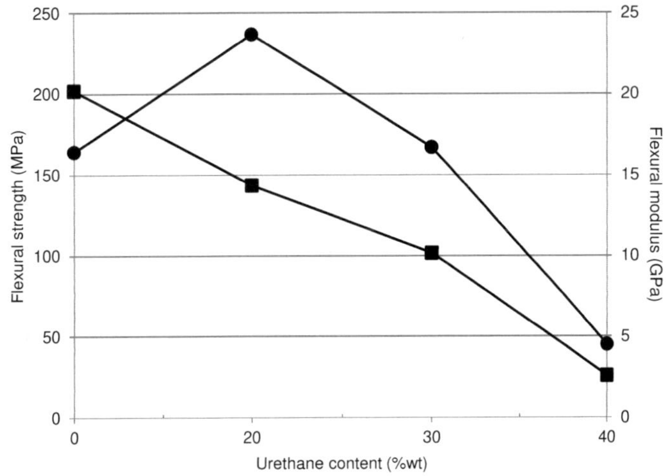

Figure 2. Flexural strength (●) and flexural modulus (■) of BA-a : PU-HDI/E polymer alloy as a function of PU-HDI/E content.

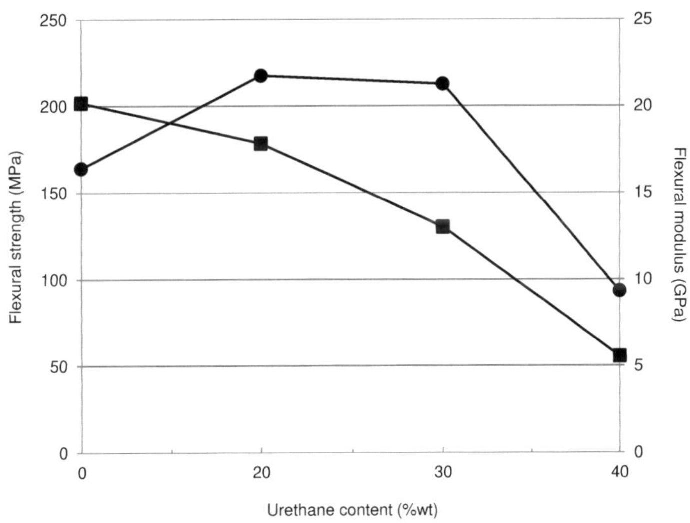

Figure 3. Flexural strength (●) and flexural modulus (■) of BA-a : PU-HDI/S polymer alloy as a function of PU-HDI/S content.

Comparison of flexural properties of the BA-a : PU alloy specimens with HDI base by varying polyol types (Polypropylene glycol; E and Poly(ethylene, diethylene adipate); S) at various urethane mass fractions exhibited in Figure 2 and 3. The results show that the reduction in flexural modulus of both figures were caused by an increase in urethane content but the relationship between urethane content and flexural strength didn't exhibit a linear behaviour. The synergistic behaviour was at the maximum with a BA : PU ratio of 80 : 20 for both polyol types. The synergistic behaviour in their flexural strengths as mentioned above was the result of the enhancement of the crosslink density of the polymer alloys with the addition of urethane.

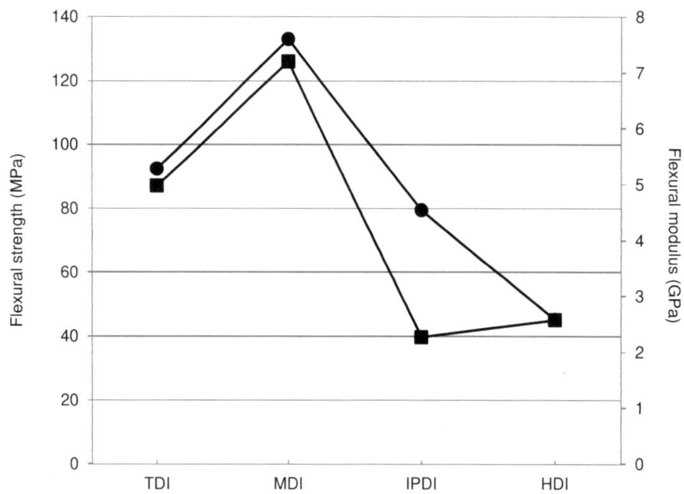

Figure 4. Flexural strength (●) and flexural modulus (■) of BA-a:PU base on Poly(ethylene, diethylene adipate) glycol (S) using different isocyanate types at component ratio 60 : 40 for polymer alloy.

Figure 4 shows the flexural properties of the BA-a : PU alloy specimens with polyester polyol base at a fixed mass ratio of 60 : 40. It is observed that maximum flexural properties were obtained when using MDI to synthesize urethane prepolymer. This statement is based on the assumption that having two benzene rings in monomeric MDI structure

5. Conclusions

In this research, alloying of BA-a : PU not only improves the thermal properties of urethane and benzoxazine but also provides further insight in shape memory research.

6. References

[1] Ionescu, M., Chemistry and Technology of Polyols for Polyurethanes, 2005.

[2] Yeganeh, H., Polybenzoxazine/Polyurethane Alloys. Handbook of Benzoxazine Resins, 2011. p. 389-402.

[3] Rimdusit, S., Jubsilp, C., and Tiptipakorn, S., Alloys and Composites of Polybenzoxazines Properties and Application, 2013. Singapore: Springer.

[4] Christian, S., Stefan, K., Rainer, S., Katharina, L., and Andreas, T., Benzoxazines for Industrial Applications Comparison with Other Resins, Formulation and Toughening Know-How, and Water-Based Dispersion Technology, 2011. p. 605-620.

[5] Rimdusit, S., Thiptipakorn, S., Jubsilp, C., and Takeichi, T., Polybenzoxazine alloys and blends: Some unique properties and applications. Reactive and Functional Polymers, 2013. 73(2): p. 369-380.

[6] Kumar, S., and Nair, R., Polybenzoxazine–new generation phenolics. Handbook of Thermoset Plastics, 2014. p. 45-73.

[7] Rimdusit, S., and Jubsilp, C., Properties Enhancement Obtained in Anhydride-Modified Polybenzoxazines. Advanced and Emerging Polybenzoxazine Science and Technology, 2017. p. 147-170.

Acknowledgements

The authors would like to thank Vencorex (Thailand) Co., Ltd., IRPC Polyol Co., Ltd. and Foamtec International Co., Ltd. for supporting us a chemical.

Effects of types of polyols on thermal degradation behaviors of poly(benzoxazine-urethane) alloys

Aumnart Klinchan[1], Panuwat Luengrojanakul[2], Sarawut Rimdusit[1*]

1,2 Polymer Engineering Laboratory, Department of Chemical Engineering, Faculty of Engineering, Chulalongkorn university, Bangkok 10330, Thailand
E-mail:sarawut.r@chula.ac.th

Abstract. In this work, thermal stability of alloys of polybenzoxazine (PBA-a) / polyurethane (PU) which show unique shape memory properties has been investigated. This study aims to confirm the thermal stability under cyclic heat activation in shape memory performance. Thermal degradation kinetics of the system is evaluated as this behaviour has not been studied before. The degradation temperatures of the obtained alloys were determined by Thermogravimetric Analyzer (TGA) with various heating rates of 15, 20 and 25°C/min. Polybenzoxazine and polyurethane alloys at weight ratios of 80:20, 70:30 and 60:40 were examined.

1. Introduction

Benzoxazine resin and urethane prepolymer are widely used in various applications, from commodities and construction materials to high-technology industries such as automotive and aerospace industry, as well as chemical industry[1]. Polybenzoxazine and polyurethane can be combined to form alloys which further enhance and broaden their intriguing properties such as high thermal stability, mechanical strength, easy processing ability, thermal degradation resistance and resistance against various solvents[2]. These alloys have also been reported to show useful shape memory behaviors under thermal activation[3]. The repeated thermal activation requires an understanding of thermal stability particularly thermal degradation kinetics of the alloys which has not been studied before.

2. Experimental section

2.1 Materials

Benzoxazine monomer was synthesized following the patented solventless synthesis technique[4]. The obtained solid monomer was then (optional) crushed into small powder and kept in a refrigerator. The urethane prepolymer was prepared by using a mixture of toluene diisocyanate (TDI) with diol (Mw = 1000) at a molar ratio of 1.5:1. Toluene diisocyanate and diol were thoroughly mixed in a distillation flask and the mixture was continuously stirred under nitrogen stream. Toluene diisocyanate and diol were thoroughly mixed via continuous stirring under nitrogen stream in a distillation flask.

IOP Conf. Series: Materials Science and Engineering **617** (2019) 012007 doi:10.1088/1757-899X/617/1/012007

2.2 Synthesis

Polybenzoxazine (PBA-a) was blended with polyurethane (PU) at various weight ratios, i.e. 80:20, 70:30, and 60:40. All of the alloy blends were step-cured in an air-circulated oven at a temperature of 150°C, 160°C, 170°C and 180°C for 2 hr, 2 hr, 3 hr, and 2.5 hr respectively.

2.3 Characterization

The decomposition temperature (T_d) of the blend were studied by using TGA instruments from Mettler-Toledo model TGA1 STARe system. TG curves are normally plotted with the mass change expressed as a percentage on the vertical axis and temperature or time on the horizontal axis. The experiments were performed from 25-850°C under nitrogen atmosphere[5]. The flow of purging nitrogen was (redundant) 50 ml/min. Alloy samples were of square shape weighing approximately 8 mg each with a dimension of 2x2 mm.

3. Measurements

3.1 Thermal gravimetric analysis (TGA)

A sample weighing approximately 8 mg was used for each test. TGA thermograms of different blends at different/various weight ratios of PBA-a/PU at various heating rates, are shown in Figure 1-3.

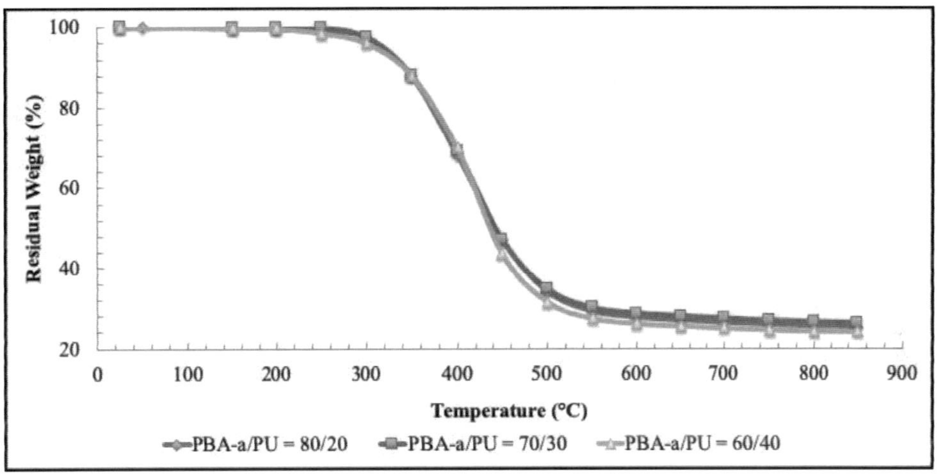

Figure 1. TGA curves of the alloys at various PBA-a/PU ratios. (Heating rate 15°C/min).

Figure 1. shows the degradation temperature of the alloys with various ratios of PBA-a/PU weight ratios and a heating rate 15°C/min. The 5% weight loss temperature for PBA-a/PU alloys of 80/20, 70/30 and 60/40 weight ratio is 319.60, 323.13, 312.12°C respectively.

IOP Conf. Series: Materials Science and Engineering **617** (2019) 012007 doi:10.1088/1757-899X/617/1/012007

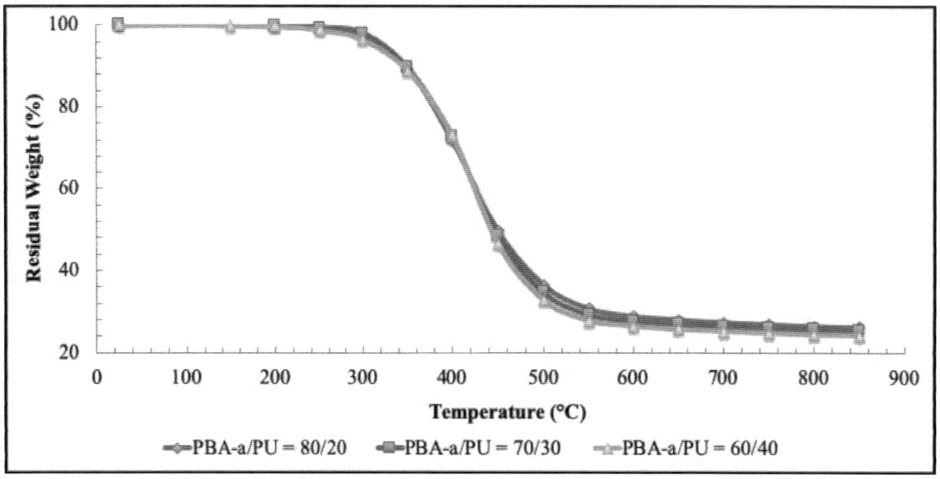

Figure 2. TGA curves of the alloys at various PBA-a/PU ratio. (Heating rate 20°C/min)

Figure 2. shows the degradation temperature of the alloys with various ratios of PBA-a/PU weight ratios and a heating rate 20°C/min. The 5% weight loss temperature for PBA-a/PU alloys of 80/20, 70/30 and 60/40 weight ratio is 327.02, 329.83, 315.93°C respectively.

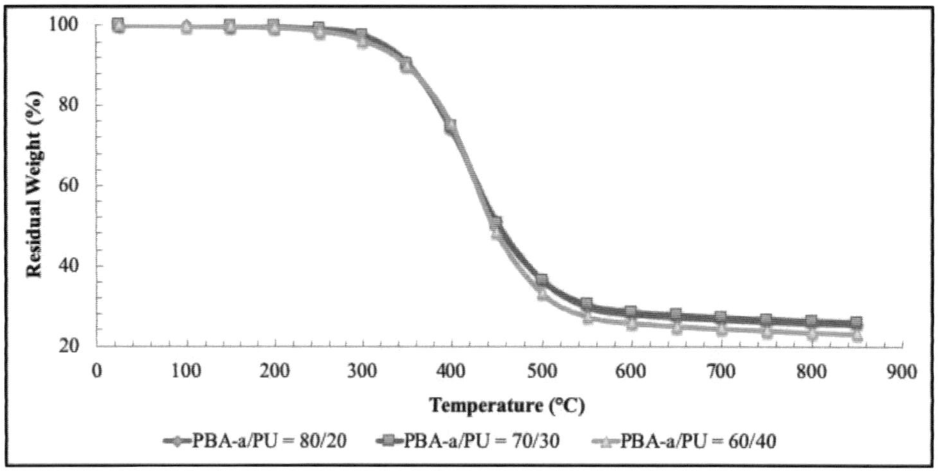

Figure 3. TGA curves of the alloys at various PBA-a/PU ratios. (Heating rate 25°C/min).

Figure 3. shows the degradation temperature of the alloys with various ratios of PBA-a/PU weight ratios and a heating rate 15°C/min. The 5% weight loss temperature for PBA-a/PU alloys of 80/20, 70/30 and 60/40 weight ratio is 327.72, 328.57, 318.26°C respectively.

Table 1. Thermal degradation temperature of the alloys at the various PBA-a/PU ratios and heating rate.

PBA-a/PU	Td @ 5% weight loss (°C)		
	Heating rate (15°C/min)	Heating rate (20°C/min)	Heating rate (25°C/min)
80/20	319.60	327.02	327.72
70/30	323.13	329.83	328.57
60/40	312.12	315.93	318.26

Table 1. shows the decomposition temperature (T_d) of the alloy blends at various PBA-a/PU weight ratios. It could be noticed that the PBA-a/PU alloy at 70/30 weight ratio provides the highest degradation temperature.

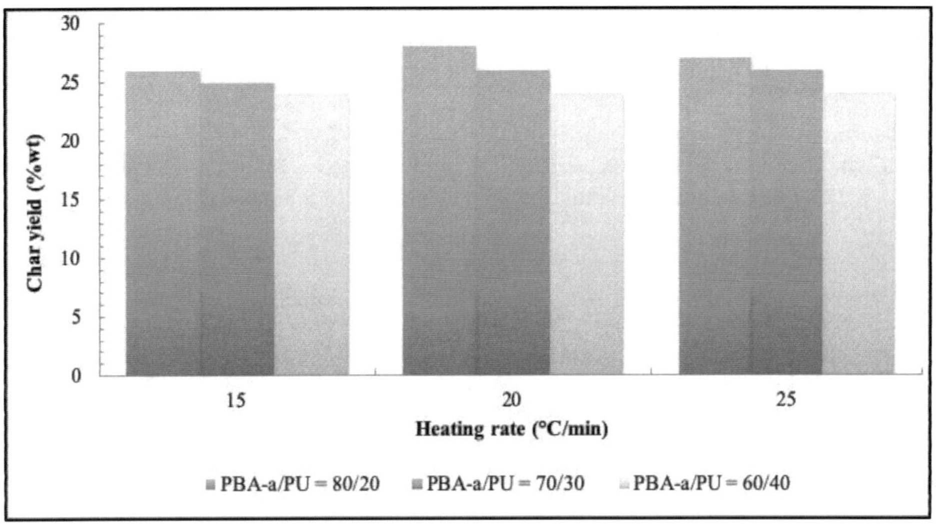

Figure 4. Char yield of PBA-a/PU blends at various PBA-a/PU ratios and temperature of 800°C.

Figure 4. shows char yield of various blends using TGA instruments. The experiments were performed using various heating rate with the temperature ranging from 25-850°C under nitrogen atmosphere and the flow purging nitrogen was kept at 50 ml/min. The sample work dimension is 2x2 mm and weighing approximately 8 mg.

4. Results and discussion

PBA-a/PU alloy of 70/30 weight ratio was determined to have the highest degradation among the various alloy blends. The 70/30 weight ratio PBA-a/PU alloy has a 5% weight loss temperature of 323.13°C, 329.83°C and 328.57°C for a heating rate of 15°C/min, 20°C/min and 25°C/min respectively. The char yield of the alloy saw an increase as the amount of PBA increases due to polybenzoxazine already having a high char yield.

5. Conclusions

From the experiment on thermal degradation of alloy blends between benzoxazine monomer and urethane prepolymer, 70/30 weight ratio PBA-a/PU alloy provided the highest degradation temperature. The result indicates the 70/30 weight ratio should be used for development due to its improved thermal properties. The work on thermal degradation kinetics of PBA-a/PU will be investigated on a further date.

6. References

[1] N.N Ghosh, B.Kiskan, Y. Yagci, Polybenzoxazine: Prog.Polym.Sci.Vol.32, (2007) 1344-1391
[2] S.Rimdusit, S.Pirstpindvong, W.Tanthapanichakoon, and S.Damrongsakkul: Polym.Eng.Sci.Vol. 45, (2005),p.288
[3] P.Prathumrat, S.Tiptipakorn and S.Rimdusit: Smart.Mater.Struct.Vol.26, (2017)
[4] H. Ishida. U.S. patent 5,543,516 (1996)
[5] S.Tiptipakorn, S.Rimdusit, P.Suwanmala, and K.Hemvichian: Advanced Materials Research. Vol. 214 , (2011),p.439-443

Acknowledgements

The authors would like to acknowledge Polymer Engineering Laboratory, Department of Chemical Engineering, Faculty of engineering, Chulalongkorn University.

Experimental description of aging of palm oil kernel shell powder/epoxy composite

Petr Valášek, Karolína Habrová, Miroslav Müller

Czech University of Life Sciences Prague, Faculty of Engineering, Department of Material Science and Manufacturing Technology, Kamýcká 129, Prague, Czech Republic
E-mail:valasekp@tf.czu.cz

Abstract. Aging of polymeric materials is one of the application limits of these materials. Describing the degradation will help determine the life cycle of the product, describing the degradation process is so important in determining possible application spheres. The paper focuses on the degradation of the matrix composite in the form of epoxy resin and palm oil kernel shells (PKS). Handling bio-waste arising from oil palm processing is a great challenge, material use of these commodities is an interesting alternative to other options. The filler was of irregular particle size up to 100 μm the filler concentration in the matrix was 2.5, 5.0 and 10.0 wt.%. Renewable resources are increasingly promising options for sustainable material engineering. The presence of organic microparticles can optimize some mechanical characteristics and reduce the cost of the resulting material. Microscopy was used to assess interphase interference and microscopic analysis of the surface and particle shape. The contribution showed that the presence of organic particles in the matrix did not significantly affect the decrease of mechanical properties due to degradation in laboratory conditions. Specifically, these were experiments in the climatic chamber where humidity and temperatures varied between + 70 ° C and -40 ° C for 840 hours. The decrease in shear strength on steel adherence reached 3.99 MPa in the unfilled resin and 5.77 MPa (10% PKS) in the palm kernel shell resin.

1. Introduction

Biocomposite materials are an interesting alternative to conventional material engineering [1]. A composite of natural fibers can be considered as a very perspective group [2]. According to Ku et al. [3], natural fibers in composite systems make materials attractive to researchers, engineers and scientists, especially due to their low cost, relatively good mechanical properties. A similar economic benefit can be expected for composites with natural particulate fillers, but the optimization of mechanical properties will be different and will interfere with other characteristics. An example may be the work of authors' colleagues [4] describing optimization especially in the field of shear strength. In composite systems, secondary raw materials arising from the processing of natural resources can also be used. An example may be palm oil, which holds the dominant global position in the production of vegetable oil.

The Asian region is one of the largest palm oil producers in the world. Both Indonesia and Malaysia are the first in the volume of production, and these countries are also responsible for handling agricultural waste generated by palm oil production. Ramli [5], in his analysis, states that production growth in Malaysia is steadily rising (by 2020 expects 5 million hectares of palm oil trees). From the point of view of the recovery of waste generated during processing, this is still a current problem. On the technological lines processing the palm oil fruits, there is an accumulation of further usable products. The first product is cellulose fibers, which can also be used in the field of material engineering. The second product is the fruit shell. They either use energy applications in different forms or can be integrated into the field of material engineering. Materials utilization in composite systems in interaction with polyester resin is described by Nabinejad et al. [6], another way to use PKS is, for example, application to rubber [7].

The experiment compares the shear strength of the overlapped bodies using PKS and the resin itself while describing the aging of the test bodies in the climatic chamber.

2. Materials and methods

In the experiment, fruits from palm oil from processing lines in Sarawak, Borneo were used (see Figure 1).

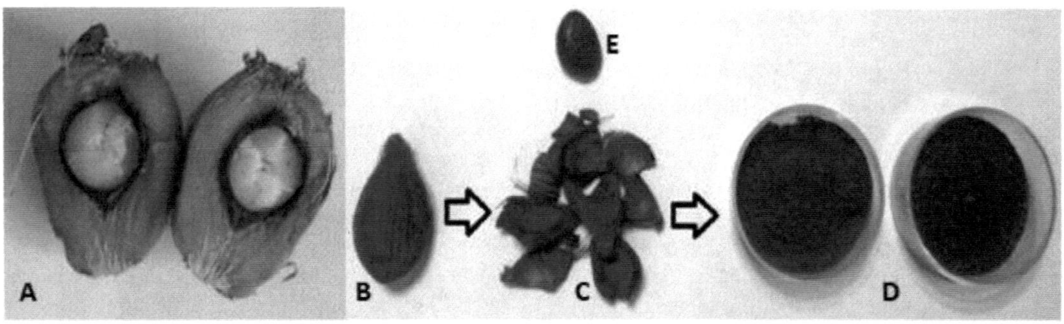

Figure 1. An illustrative example of oil palm fruit handling: A-fruit after harvest, B-fruit processed on a technological line (shell and core), C-shells after core removal, D-prepared particles, E-core, that is pressed on technological line.

Shells were processed with a knife mill at a speed of 20,000 rpm·min^{-1}, after which the particles were dried in an oven at 105 °C. Subsequently, individual fractions of PKS were prepared through analytical sieves. After the preparation of the individual fractions, composite systems were made such that the epoxy resin was mechanically mixed with PKS. The resulting concentration of PKS in the matrix corresponded to 2.5, 5.0 and 10.0 wt.%.

For the description of shear strength on steel sheet S235J0, the standard CSN EN 1465 was used. The roughness parameters of the steel sheets prior to application of the composite mixture were measured by the touch grinder. The conditions defined in CSN EN ISO 914 were used to simulate the aging of the material in the climatic chamber. The total degradation time was 840 hours, which is 35 cycles. One cycle of degradation corresponded to 24 h. In one cycle, the bodies were tempered to 70 °C at 90% humidity for 16 hours, then subcooled to -40° C for 3 hours and then returned to + 70 °C at 50% humidity.

Electron microscopy was used for the optical analysis (Tescan Mira 3 GXM), for the evaluation was used ANOVA.

3. Results and discussion

Optical analysis of particle size and particle shape was performed using microscopy, and the surface of the particles was then analyzed on an electron microscope (Figure 1-a). The particle surface (Figure 2-b, c) is covered by pores that are responsible for increased water absorption [8]. It should also be noted that the surface of the shell is covered with fibers, according to [9] the fibers are up to 30% of the surface, as can be seen from Figure 1 c. After the optical analysis, the particle size was determined as follows: 58 ± 29 μm. Optical surface analysis of PKS confirmed that microscopic structure even with the presence of pores is very similar to that of other types of fruit shells, e.g. coconuts.

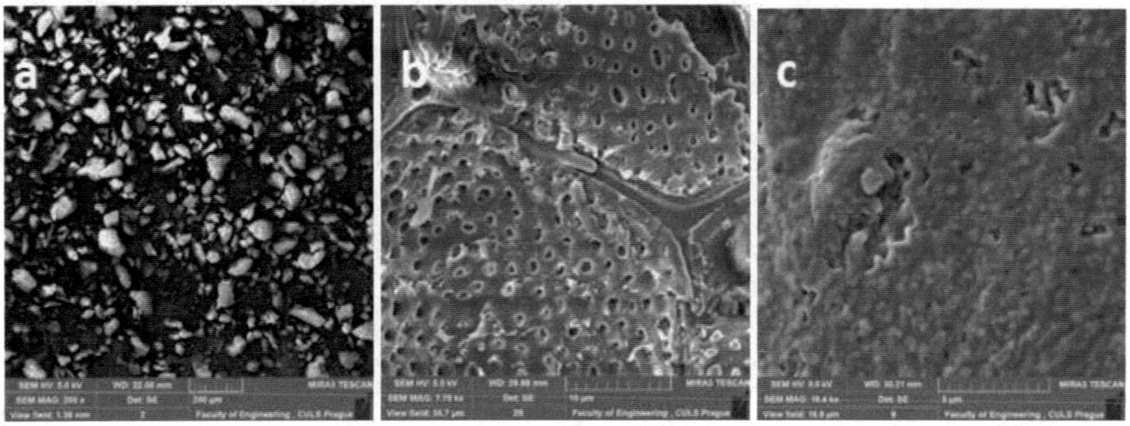

Figure 2. a-Overview picture of PKS with size up to 100 µm, b-surface of PKS with visible pores, c-detail of PKS surface.

The shear strength is influenced by the roughness of the surface of the material (Ra = 2.7 ± 0.6 µm and Rz = 11.9 ± 3.2 µm). The shear strength of the resin without filler was 11.75 ± 0.85 MPa. With the addition of filler, shear strength was increased by 1.18 MPa (2.5% PKS), 0.70 MPa (5.0% PKS) and 1.13 MPa (10.0% PKS) - adhesive type of failure. An important aspect affecting the resulting mechanical properties of composite systems is the interfacial interaction. In the case of oilseeds, it is necessary to point to a higher proportion of oils in plant residues, which can adversely affect the formation of interfacial phases. For this reason, the mutual phase interaction was carried out by optical analysis (see Figure 3), which confirmed the corresponding interfacial interaction.

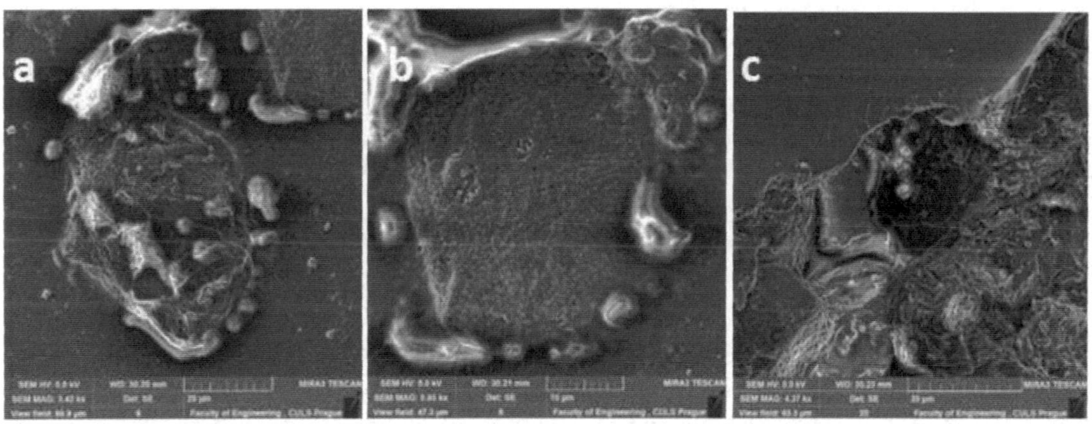

Figure 3. Description of interfacial interaction by electron microscopy: a-Mag. 3.42 kx, b-Mag. 5.35 kx, d-Mag. 4.37 kx.

The course of degradation compared to the filled and non-filled resin is shown in Figure 4. The presence of PKS fillers did not significantly increase the rate of degradation. After the degradation cycle (840 h), the non-filler resin strength dropped to 7.76 ± 0.83 MPa, PBS composite with 2.5% filler to 7.93 ± 1.21 MPa, 5.0% to 7.98 ± 1.15 MPa and 10.0% to 7.52 ± 1.46 MPa - adhesive type of failure. Conclusions of the experiment confirm the possible use of agricultural waste in the form of material engineering, as described by a number of authors [2, 8, 10, 11].

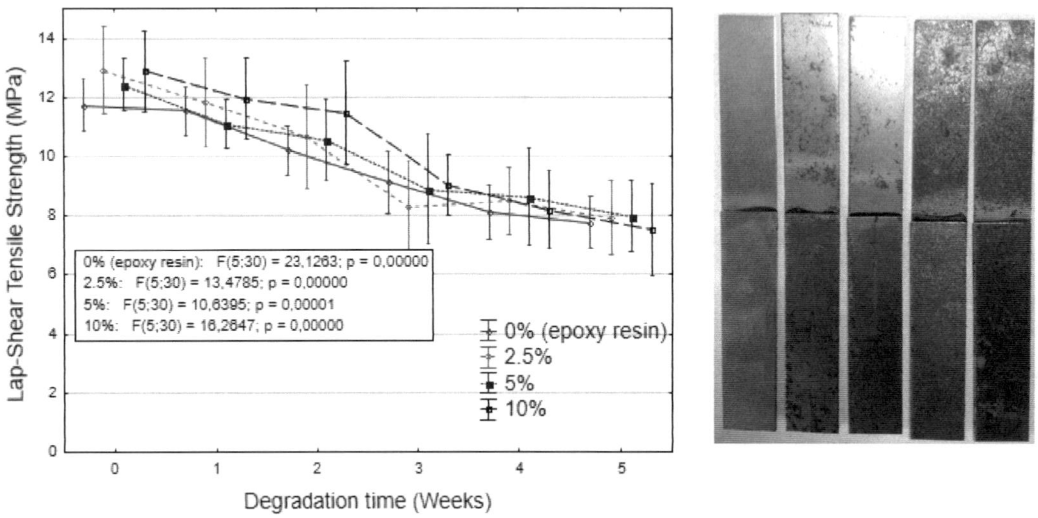

Figure 4. Course of degradation of the non-filled and filled resin (interval 0-840 hours - left), the test body before degradation and corrosion of the steel body after the degradation - after 840 hours (right).

4. Summary

The article describes a possible alternative of the use of PKS, which could contribute to measures leading to the exploitation of by-products of agriculture - sustainable development. It is an environmentally friendly way of using materials to prevent environmental pollution and minimize the consumption of primary raw materials. The disadvantage of the biological microparticles used may be an increased proportion of vegetable oil in these agricultural residues and a tendency to absorb water. The performed experiment did not show an increase in the rate of degradation of the test bodies with PKS comparing to the unfulfilled resin. At the same time, the experiment has shown that the presence of PKS slightly increases the shear strength of the assemblies. Optical electron microscopy analysis confirmed the good wetting of PKS with epoxy resin.

5. References

[1] P.A. Fowler, R.M. Elias: Technology, environmental credentials and market forces: Journal of the Science of Food and Agriculture Vol. 86(12/2006), pp. 1781–1789.

[2] A. Ruggiero, P. Valášek, M. Müller. Exploitation of waste date seeds of Phoenix dactylifera in form of polymeric particle biocomposite: Investigation on adhesion, cohesion and wear: Composites Part B: Eng Vol. 104 (2016), pp. 9–16.

[3] H. Ku, H. Wang, N. Pattarachaiyakoop, M. Trada: A review on the tensile properties of natural fiber reinforced polymer composites. Composites: Part B 42 (2011), pp. 856– 873.

[4] P. Valášek, K. Habrová, M. Müller: Degradation of strength properties of epoxy resin filled with natural-based particles: Vol. 106 (2018), Code 136927.

[5] A. Ramli: Short-term and long-term projection of malaysian palm oil production: Malaysian Palm Oil Board (MPOB), Vol. 3 (2003), pp. 32–36.

[6] O. Nabinejad, D. Sujan, M. E. Rahman, Ian J. Davies: Effect of oil palm shell powder on the mechanical performance and thermal stability of polyester composites: Materials & Design (1980-2015), Vol. 65(2015), pp. 823–830.

[7] D.Shuhairiah, I. Hanafi, A.B. Azhar: The Effect of 3-aminopropyltrimethyoxysilane (AMEO) as a Coupling Agent on Curing and Mechanical Properties of Natural Rubber/Palm Kernel Shell Powder Composites. Procedia Chemistry, Vol. 19 (2016), pp. 327–334.

[8] U.J. Alengaram, H. Mahmud, M.Z. Jumaat: Development of lightweight concrete using industrial waste material, palm kernel shell as lightweight aggregate and its properties. In:

2nd International conference on chemical, biological and environmental engineering (2010), pp. 277–281.
[9] Agus SetyoMuntohar, Muhammad EkhlasurRahman: Lightweight masonry block from oil palm kernel shell: Construction and Building Materials, Vol. 54 (2014), pp. 477- 484.
[10] P. Valášek, M. Müller: Tensile characteristics of epoxy/jute biocomposites prepared by vacuum infusion: Lecture Notes in Electrical Engineering Vol. 505 (2019), pp. 574-548.
[11] P. Valášek, K.H. Ambarita: Material usage of oil-palm empty fruit bunch (EFB) in polymer composite systems: Manufacturing Technology Vol. 5 18 (2018), pp. 686-691.

Acknowledgments
This paper has been made with the assistance of the grant IGA TF CZU.

IOP Conf. Series: Materials Science and Engineering **617** (2019) 012010 doi:10.1088/1757-899X/617/1/012010

Weatherability Temperature Field and Temperature Stress Simulation of Inorganic Mortar Compound Panel with Insulation and Decoration Based on Actual Engineering Model

Hu Xiang-lei[1], Wang Lin[2]

1 College of Architecture and Urban Planning, Tongji University, China
2 Tongji Architectural Design（Group）Co., Ltd., China
E-mail：huxianglei@163.com

Abstract. Finite element simulation software can explore the heat transfer and deformation of complex shapes and can be used as an auxiliary tool for large weathering tests. Based on the results by ANSYS software, this paper explored temperature field and temperature stress field of inorganic mortar compound panel with insulation and decoration (IMCP), considering weather resistance experiment based on practical engineering. The paper holds that the relevant regulations should provide the mortar adhesion area reinforcement of IMCP may be not less than 80%. Moreover, in order to adapt to the temperature deformation, the width of the joint of IMCP should also be specified and recommended not be less than 5 mm.

1. Introduction

The Inorganic Mortar Compound Panel with Insulation and Decoration (IMCP) is platy product, which is prefabricated in the factory and recombined by inorganic mortar insulation material, decorative panels, adhesives and connecting parts, with the functions of insulation and decoration and excellent fireproof performance. As progressive measure of industrial assembly technology, insulation decorative panels meet the technological trends. The weathering test of building materials is an encryption combination of artificial simulation of high temperature rainfall and thermal-cold cycle. Practice has proved that large-scale weathering test is closely related to wall material safety and durability. Weatherability is one of the issues to be studied in IMCP, but large-scale weathering test requires more stringent test conditions and longer time. The finite element numerical simulation, intuitive and convenient, can get a temperature filed close to any shape, any time and consider the influence of complex shape on temperature distribution. ANSYS, the general-purpose finite element numerical simulation software, enables data sharing and exchange with most computer-aided design software, and be widely used in railway, petrochemical engineering, machinery manufacturing and other fields. According to the pilot project model, this payer has simulated the local heat transfer of IMCP, analysed the temperature field and temperature stress under the weathering test conditions through ANSYS model building.

2. Simulation analysis of local heat transfer

IOP Conf. Series: Materials Science and Engineering **617** (2019) 012010 doi:10.1088/1757-899X/617/1/012010

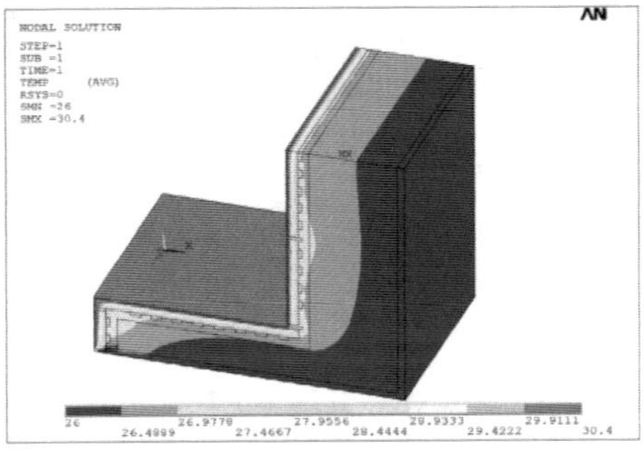

Figure 1. Temperature profile of 40 mm IMCP in summer.

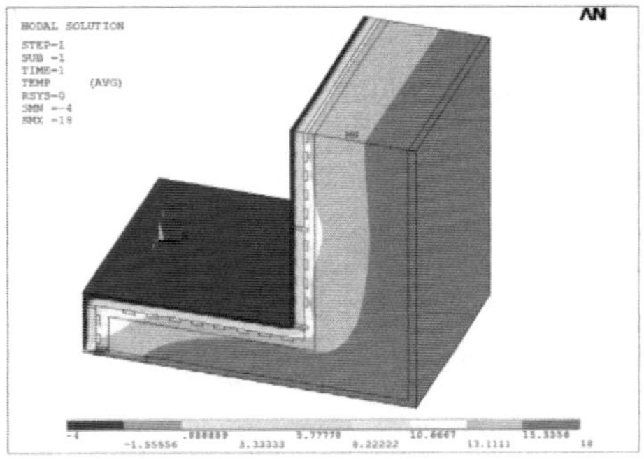

Figure 2. Temperature profile of 40 mm IMCP in winter.

IMCP is composed of the panel, the fixing member and adhesive mortar. By analysis of fixing members at the composite panel wall surface and intersection portions of convex windows, and effect of air layer on heat transfer, this simulation provides a basis for simplification of wall models under the weathering conditions, increases the computational feasibility, shortens the calculation time. The calculation model consists of external ambient temperature, composite panels, fixing members, adhesive mortar, plastering mortar, building walls and indoor ambient temperature elements. The model mesh is divided into five layers from the outside to the inside, which are respectively the insulation decorative composite panel, the fixing members, the air, the adhesive mortar layer, the wall and the fixing members located between composite panel and wall. The software simulates the heat exchange of IMCP with thickness 40 mm and 60 mm in winter and summer. The outdoor temperature is 30.8 °C in summer, -2.0 °C in winter, and the indoor temperature is 26 °C in summer and 18 °C in winter. Physical properties are assigned via software. As shown from the wall temperature distribution Figure 1 obtained from simulation, the metal fastenings are weak parts of insulation, the temperature of the part of contact with the wall is lower than that of other parts, but the material insulation performance is not significantly affected; although the air layer in the mortar interval is beneficial to the insulation, the heat insulation performance is not increased significantly because of the thin thickness. According to Figure 2, the heat transfer direction of the wall in summer is opposite, but the temperature curve is similar to that in winter. The metal fastening does not significantly affect the thermal insulation performance of the material, and the air layer in the adhesive paste is also insignificant for improving the thermal insulation performance. The temperature profile of 60 mm IMCP is similar to 40 mm. The metal fastening has no significant effect on the heat transfer of the wall, and the effect of the air layer is also limited.

Figure 3. Pilot high-rise residence.

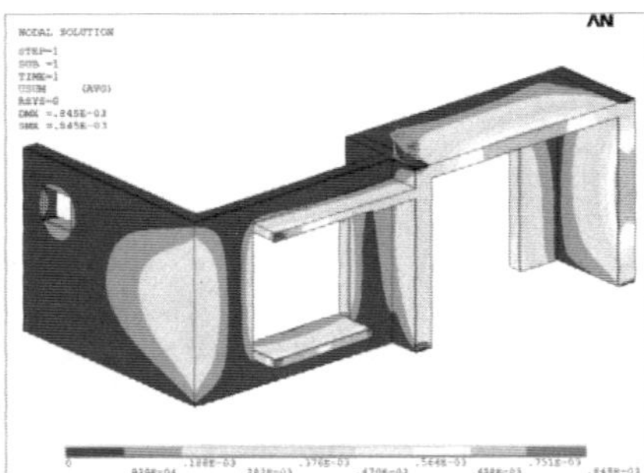

Figure 4. Preliminary analysis of thermal deformation of IMCP.

3. Numerical simulation under weathering conditions

The Industry Standard of the PRC Code, Materials of External Thermal Insulation Systems Based on Insulation Decorative Panel (JG/ T 287 — 2013), has stated clearly the test samples and test steps of large weathering test. The samples are made of concrete walls and tested external insulation system, with about 0.5 m^2 concrete wall used as the base layer and 0.24 m^2 window opening on the wall. The test steps include thermal rain cycle, thermal-cold cycle, freeze-thaw cycle and tensile bond strength test. After analysis of the external insulation system weathering test and ANSYS simulation, Chinese scholar concluded that the results of weathering simulation and test results have good correlation [1]. In this paper, part of standard floors of high residence was selected, and ANSYS is used to simulate the heat rain cycle, thermal-cold cycle, freeze-thaw cycle on temperature field and temperature stress. Figure 1 and Figure 2 show that the temperature of the wall substantially changes along the thickness direction of the system itself, and therefore the simulation assumes that the materials in each layer is uniform, continuous and isotropic, the material layer is connected tightly, inter-layer thermal resistance is ignored, the insulating layer joint is not considered and thought to be continuous and the anchor bolt thermal bridge is not considered. Performance parameters and thickness of the materials of each layer are mentioned in Table 1. The software analysis process involves building finite element model, applying loads, solving and post processing. Firstly, preliminary analysis of the local models of the pilot building including the outer wall is carried out (Figure 3, Figure 4). The sizes of one outer wall and window holes follow the requirements of large weathering test samples. Figure 4 shows the balcony is used as an outer wall member, and its temperature field and temperature stress change are similar to those of bay windows. To simplify the computational effort, the balcony is cancelled, the bay window is retained and the cell density is manually controlled.

Table 1. Performance parameter.

Material parameters	Density / [kg/m^3]	Thermal conductivity / [W /(m · K)]	Specific heat capacity [J /(kg · K)]	Modulus of elasticity / [Pa]	Poisson's ratio	Thermal expansion coefficient / [°C]	Thicknes s/ [mm]
RC	2300	1. 74	920	20.0×10^9	0. 20	1.0×10^{-5}	200
Mortar creed -coat	1600	0. 93	1050	4.9×10^9	0. 28	1.2×10^{-5}	20
Bonding mortar	1800	0. 93	1050	4.9×10^9	0. 28	1.2×10^{-5}	5
IMCP	160	0. 05	1550	17×10^9	0. 18	1.4×10^{-5}	40

The boundary conditions in contact with the exterior wall of the wall are simplified according to the steps of the weathering test. The IMCP is positioned in the test chamber, and the convective heat transfer coefficient in the chamber environment is taken as 8.7 W/(m^2·K) (at watering cycle stage, 3000 W/(m^2·K)is taken); the convective heat transfer coefficient of the interior facing of the wall is the same as that of the exterior facing of the IMCP [2]. Due to the high internal sealing of the test chamber, there is only the change in ambient temperature, and the effect of radiative heat transfer on heat transfer is ignored.

For the purpose of the temperature field calculation of the wall, the internal temperature of the wall is first determined at $t = 0$. Different initial conditions have greater influence on calculation results for the first few cycles, and the influence will become increasingly smaller after several cycles. The temperature field and temperature variation regularity of each structural layer of the wall can be obtained through cycles, the amount of data and the amount of calculation is large, so initial conditions of weathering test in "Materials of External Thermal Insulation Systems Based on Insulation Decorative Panel(JG/ T 287 — 2013)" are simplified: 1) Within 4-6h of thermal rain cycle, the temperature inside and outside the wall is the same, and after 2 h, wall temperature at each point is basically near 25 °C. Assume that the temperature in the wall is equal at the initial time of each cycle, and after the initial node temperature is set to 25 °C, 1 thermal rain cycle is simulated. 2) From

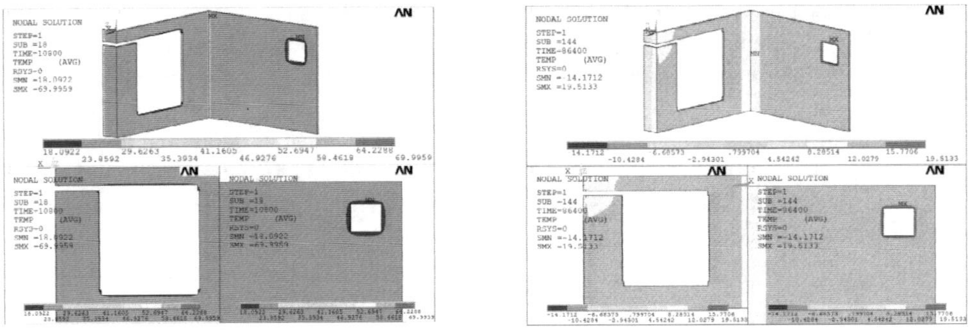

a. Thermal rain cycle b. Thermal-cold cycle

Figure 5. Inner surface temperature field of 40 mm IMCP.

the 10 h to 24h of the thermal-cold cycle, the wall temperature is always maintained at – 15 °C, and the indoor temperature is always maintained at 25 °C; assume that steady state heat transfer is reached after 14h, the temperature at each node in the wall at the initial time of each cycle is obtained, and one thermal-cold cycle simulation is conducted. 3) The freeze-thaw cycle follows the same rules as the thermal rain cycle and the thermal-cold cycle. The temperature at each node in the wall is obtained by steady-state analysis at the initial time of each cycle, and a cycle simulation is performed [3].

The wall temperature field of thermal rain cycle, thermal-cold cycle and freeze-thaw cycle of 30 mm, 40 mm and 50 mm thick IMCP is obtained through simulation. Due to the large number of load steps, the internal surface temperature field of the insulating layer at the thermal rain cycle time point 10800 s, thermal-cold cycle time point 86400 s, and freeze-thaw cycle time point 25200 s is selected, and the different colours correspond to the respective temperature bands. Figure 5 is the internal surface temperature field of 40 mm thick IMCP. According to the temperature field simulation diagram, the following conclusions can be drawn: 1) When the thickness of the insulating layer is the same, the temperatures at nodes of the inner surface of the insulating layer away from the window are almost the same; the window part is affected by the hole, and the window edge isotherm gradually decreases from the outside to the inside. 2) As the thickness of the insulating layer increases, the heat-insulation capacity of the wall gradually increases, and the amount of increase gradually decreases. Take the thermal rain cycle as an example, when the thickness of the insulating layer is 30 mm, 40 mm, and 50 mm respectively, the temperature of the inner surface of the insulating layer away from the window is 24.5701 °C, 23.8592 °C and 23.5589 °C respectively, and the heat insulation effect is obvious. 3) The thickness variation of the insulating layer has little effect on the diffraction range of the hole temperature. Through the time history post-processing of the simulation results, the thermal insulation effect of the entire circulation process of the insulating layer can be visually displayed. In this paper, five nodes are selected from 40 mm thick complex panel, i.e. the node 1 on the outer surface of the insulating layer, node 2 on the inner surface of the insulating layer, node 3 on the outer surface of the levelling layer, node 4 on the inner surface of the levelling layer, and node 5 on the concrete surface (Figure 6). According to the temperature curve, the following conclusions are drawn: 1) the nodes at the window are susceptible to the ambient temperature. The thermal insulation effect of the adhesive mortar is very limited, and the temperature change curve is very close to the temperature change of the wall structure; 2) because the convection coefficient is small and the heating time is too short, the node temperature of the outer decorative surface and outer surface of the insulating layer cannot reach the ambient temperature quickly, and at the heating section (or cooling section) and the constant temperature section, the temperature curve is always in the rising (or falling)

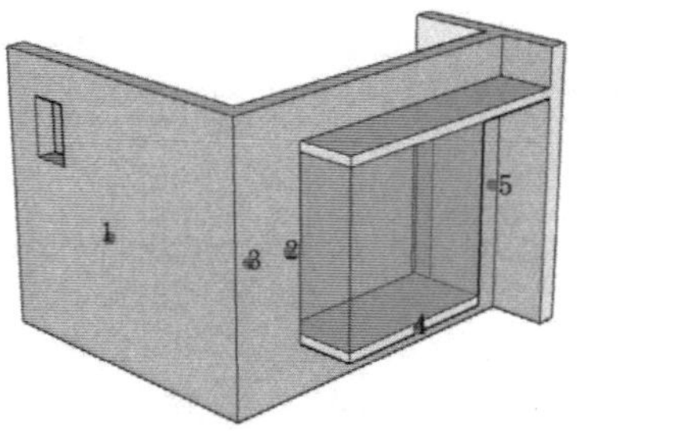

1- node on the outer surface of the insulating layer; 2- node on the inner surface of the insulating layer; 3 - node on the outer surface of the levelling layer; 4- node on the inner surface of the levelling layer; 5- node on the concrete surface.

a. Sampling position

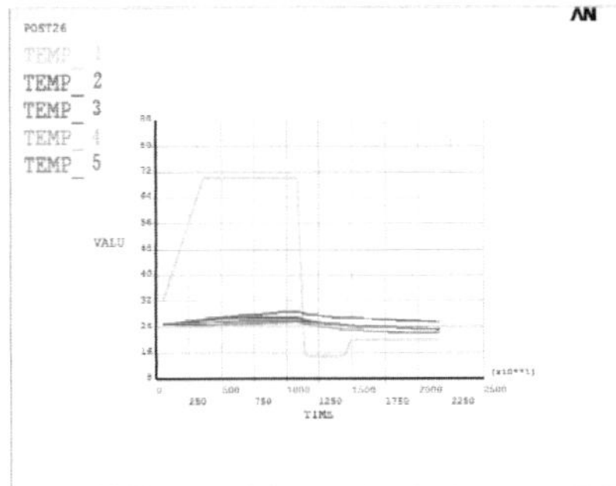

b. Thermal-rain cycle

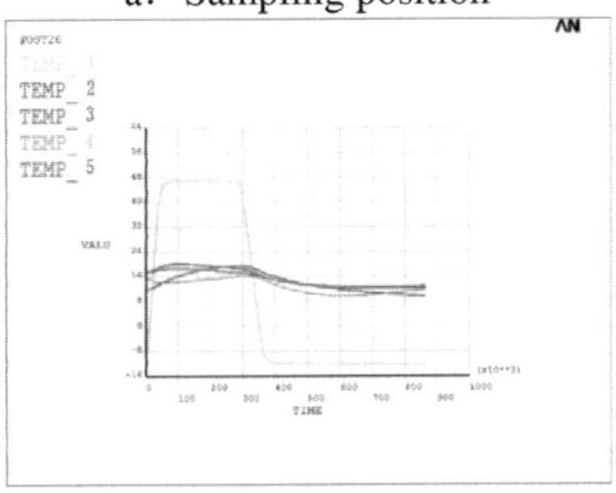

c. Thermal-cold cycle

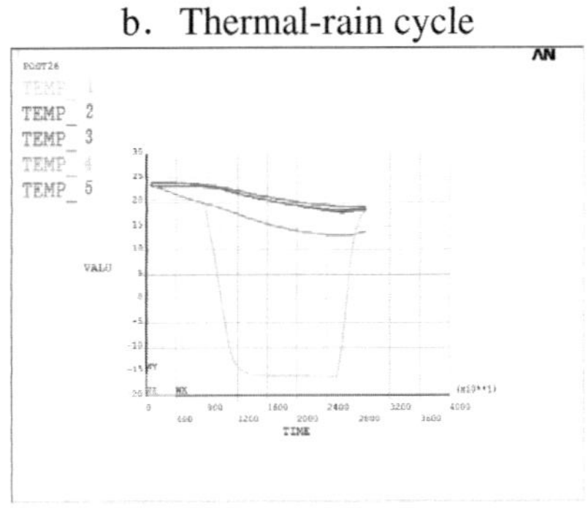

d. Freeze-thaw cycle

Figure 6. Temperature change curve of 5 nodes.

stage; 3) the thickness variation of the insulating layer has little effect on the temperature diffraction range caused by the hole. As the thickness of the insulating layer increases, the temperature difference between the inner and outer surfaces of the insulating layer and the inner and outer surfaces of the entire external insulation system gradually increases; 4) the insulation effect of the external insulation system is more obvious. The node temperature of concrete inner surface fluctuates, about 8 °C for thermal rain cycle and thermal-cold cycle, and about 10 °C for freeze-thaw cycle.

4. Numerical simulation of wall temperature stress under weathering conditions

Bonding mortar plays an important role in the connection of IMCP and wall. The difference in thermal expansion coefficients causes a large temperature stress in the adhesive mortar when the temperature changes. When the temperature stress is larger than the adhesive force, the safety and reliability of the system will be affected aversely. Through the ANSYS numerical simulation of the IMCP system of different thickness and different position, the temperature stress distribution regularities of adhesive mortar under the weathering conditions can be analysed. In this paper, the indirect method in ANSYS 10.0 thermal stress is used to conduct simulation solution of the thermal stress of the external thermal insulation system. The steps include establishing a finite element model, applying a temperature load, adding material performance parameters, applying loads and solving and post processing. The thermal stress and displacement under different insulation thicknesses are mainly post-processed.

When temperature load is applied, the boundary conditions and initial conditions in contact with the facing are the same as the simulation of temperature field. The greater the temperature difference between the inside and outside of the wall, the greater the temperature stress. The simulation analysis uses the stress and displacement under the maximum temperature difference during the cycle. The thermal rain cycle temperature is 70 °C, time point is 3750 s, thermal-cold cycle is 50 °C, time point is 1200 s, freeze-thaw cycle is - 20 °C and time point is 1650s. The original temperature of the wall temperature field is taken as 25°C. Figure 6 shows the temperature field of the internal surface of thermal rain cycle insulating layer of 40mm IMCP. Through the thermal coupling simulation under the weathering test conditions, the stress distribution and displacement distribution of the insulating layer and the adhesive mortar show the following characteristics: 1) Stress distribution regularity and displacement distribution regularity of IMCPs and adhesive mortar under the thermal rain cycle, thermal-cold cycle and freeze-thaw cycle are basically the same. 2) When the thickness of the insulating layer is the same, the temperature stress on the inner and outer surfaces of the adhesive mortar is similar to that on the window portion (excluding the boundary); the window portion is sharply enlarged, the window corner reaches the maximum, and becomes smaller outward. As the thickness of the insulating layer increases, the temperature stress experienced by the adhesive mortar layer decreases. 3) When the thickness of the insulating layer is the same, the displacement at the window and the corner is significantly increased, but the displacement at the portion away from the window and the corner is small. The thermal conductivity of the window portion changes greatly, the temperature changes drastically, and the displacement increases significantly. The corner portion undergoes a larger displacement due to geometric discontinuities and weak constraint. As the thickness of the insulating layer increases, the range of influence of the window on temperature stress and deformation becomes larger. Major factors influencing the temperature stress simulation are complicated and related to the factors such as elastic modulus, Poisson's ratio and thermal expansion coefficient of the material, and correlative factor coupling relationship has to be further explored.

5. Conclusion

Finite element simulation software can easily explore the heat transfer and deformation of complex shapes, can be used as an auxiliary tool for large weathering tests. On the basis of the results of ANSYS software analysis, external corners of walls, door and window openings and surroundings are the wall temperature stress concentration zones. To reduce the effect of temperature stress, it is recommended that the relevant regulations should provide that the mortar adhesion area reinforcement of IMCP may be not less than 80%. Moreover, in order to adapt to the temperature deformation, the width of the joint of IMCP should also be specified, and recommended not be less than 5 mm.

6. References

[1] Bo H. 2009 *Study on durability and evaluation of external insulation system for building exterior wall* (Wuhan: Huazhong University of Science and Technology)
[2] Huang Z and Tu F 2011 *Technological theory and application of external insulation* (Beijing: China Building Industry Press)
[3] Liu D. 2014 *M 2014 Housing Science 9 26–28*

Acknowledgements

This paper is supported by Open Projects Fund of Key Laboratory of Ecology and Energy-saving Study of Dense Habitat（Tongji University）, Ministry of Education. (Grant No. 201820103).

IOP Conf. Series: Materials Science and Engineering **617** (2019) 012011 doi:10.1088/1757-899X/617/1/012011

The preparation and characterization of cellulose derivative membrane as matrix for sustained release formulation

Saptono Hadi[1*], Budi Hastuti[2], and Riskita Suryaningtyas[2]

[1] Department of Pharmacy, Faculty of Mathematics and Natural Sciences, Universitas Sebelas Maret, Surakarta, Indonesia
[2] Department of Chemistry Education, Faculty of Teacher Training and Education, Universitas Sebelas Maret, Surakarta, Indonesia
*E-mail: saptono.hadi@staff.uns.ac.id

Abstract: A synthesis of cellulose derivative membrane has been performed as good candidate for sustained release matrix. Alginate is a polysaccharide that water-soluble, consisting of β-D-manuronate and α-L-guluronate, and contains of carboxylic (–COOH) and hydroxyl (–OH) groups which allow to form hydrogen bonds with drug compounds. The purpose of this study is to determine the physicochemical properties of alginate membranes. The membranes were prepared by the solution casting method on polypropylene container. Structural characterization of the membrane was done by fourier transform infrared spectra (FTIR) and the mechanical properties of the membrane was analyzed with tensile strength test and elongation test based on variation of alginate concentration. Scanning Electron Microscopy (SEM) was done to determine the surface morphology of the membrane. The FTIR spectra showed the presence of carbonyl group (C=O) at 1597 cm^{-1} and hydroxyl group (–OH) at 3426 cm^{-1}. The tensile strength of membrane for alginate with different concentration at 3.3, 5, and 10% w/v were 8.84, 17.07, and 7.36 MPa, respectively. This indicates that the higher concentration of alginate membrane will increase the strength of the membrane, but at higher alginate concentration the membrane strength tend to decrease.

Keywords: Alginate, membrane, sustained release formulation

1. Introduction

The use of thin films from natural polymers tend to increase and expand in various fields such as the pharmaceutical, medical, biotechnology, environmental, and agricultural industries [1-3]. Non-porous polymer films are also widely applied as membranes in the pervaporation process. This is due to the superiority of non-porous film layers that offer high permeability, mechanical strength, and selectivity, and their ability to separate azeotropic solutions [4].

Film material is one of the variables that greatly determine the film performance. Natural polymers are currently receiving attention from researchers in film synthesis as they are low-cost materials and also their nontoxic, biodegradable, and biocompatible properties [5]. Alginate (Figure 1) is a linear polysaccharide consisting of β-D-manuronic acid monomers and α-L-glucoronic which are connected through a 1.4 bond. Alginate is very important because of its extensive use in various industrial fields [6]. Alginates which are polyanionic when dissolved in the right conditions can interact with each other through carboxyl groups of alginate [7,8]. Alginate is also a linear polysaccharide extracted from brown algae, it dissolves in aqueous solutions. It contains a variety of 1-4 linked α-L-glucoronic and β-D-mannuronic acid residues [9,10].

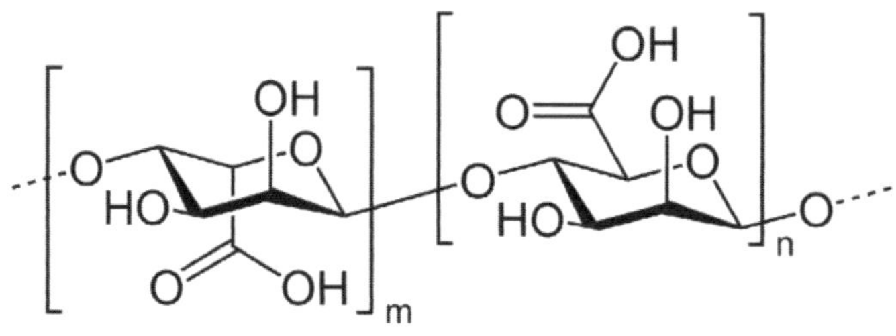

Figure 1. Chemical structure of alginate

In this study, preparation and characterization of membranes formed by alginate at various concentrations was performed. Membrane will be applied as a slow-release matrix of drugs. Slow release is a form of method that is designed to slowly release the drug into the body or gradually to release it longer, prolong the action of the drug, maintain continuous levels of drug therapeutic range and improve patient compliance. Structural characterization of the membrane was done by FTIR and the mechanical properties of the membrane was analyzed with tensile strength test and elongation test based on variation of alginate concentration. SEM test was done to determine the surface morphology of the membrane.

2. Methodology

2.1 Material
Alginat (food grade) was obtained from CV Nura Jaya Surabaya, aquadest

2.2 Membrane preparation.
Alginate was dissolved in water with variation concentration of 3.3%, 5%, and 10% w/v and were stirred until homogeneous and then printed in a polypropylene container. The solution was left for about 5 minutes and then dried at 60 °C for 24 hours to form a thin layer of membrane.

2.3 Membrane characterization.
Structural characterization of the membrane was done by Fourier transform infrared spectra (FTIR) and the mechanical properties of the membrane was analyzed with tensile strength, and SEM test to determine the surface morphology of the membrane.

3. Results and discussion

3.1 FTIR spectra
FTIR test is an observation of functional groups contained in the membrane. From Figure 2, it can be seen that the C–O vibration peak which is characteristic of the saccharide structure is located at a wavelength of about 1004-1300 cm^{-1}. The membrane has a new vibration peak of carboxyl salts at a wavelength of 1597 and 1419 cm^{-1} which shows the C=O group. The peak at a wavelength of 3410 cm^{-1} indicates the –OH group.

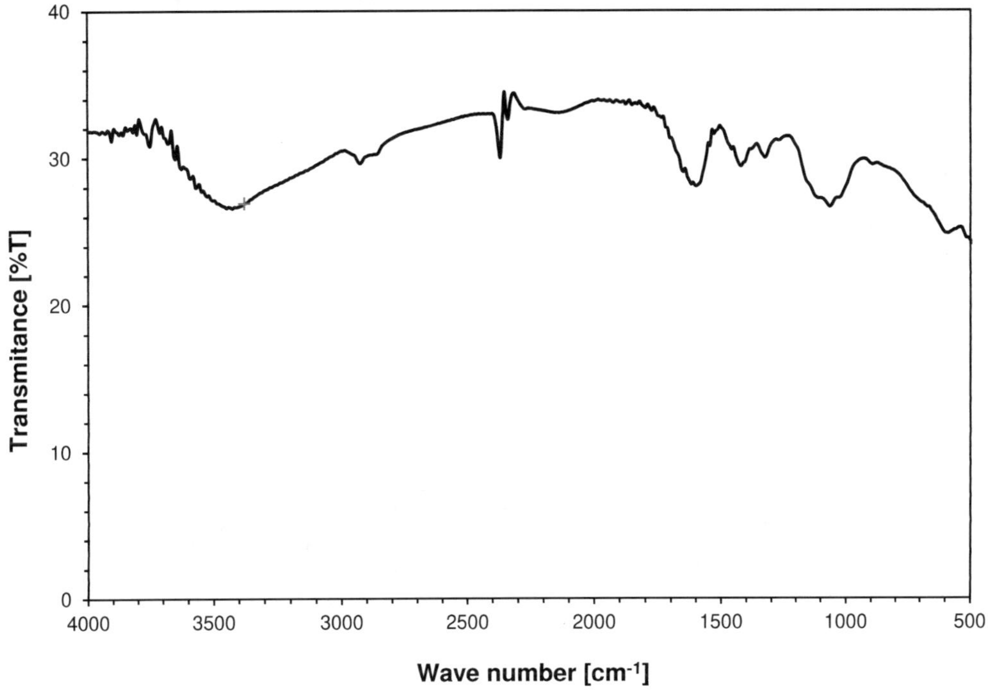

Figure 2. FTIR spectra of alginate membrane under study

3.2

XRD spectra

diAnalysis with XRD spectrometry was intended to determine the crystallinity of alginate. The ffractogram is shown in Figure 3.

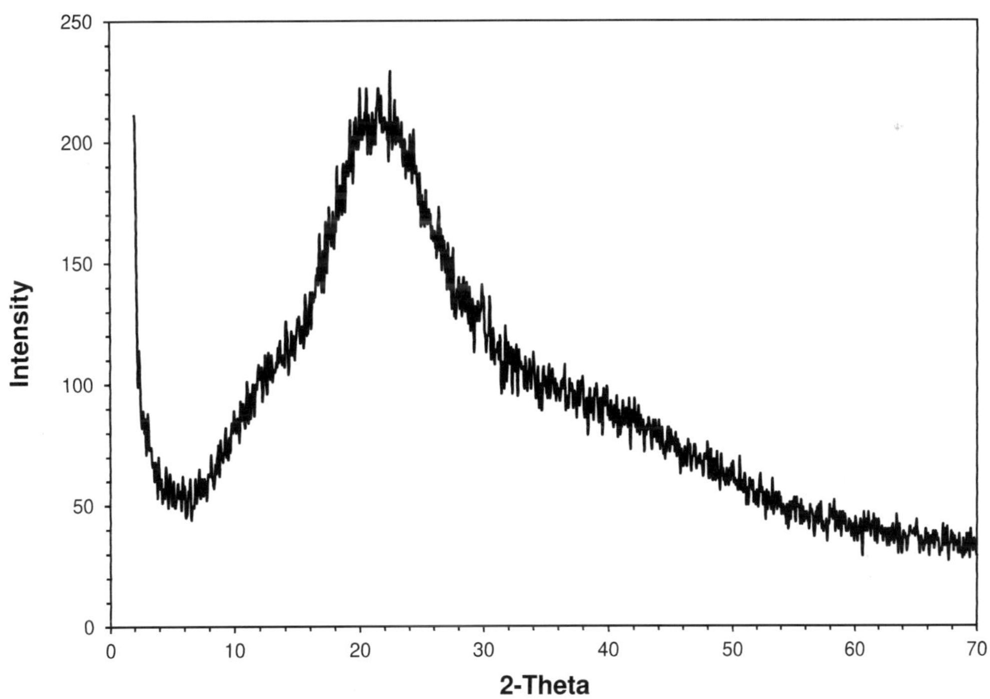

Figure 3. XRD chromatogram of alginate membrane under study

The diffractogram of alginate showed slightly sharp peak at $2\theta = 25°$, indicating semi-crystalline characteristic of alginate. The crystal structure of the alginate is strongly influenced by the hydrogen bonds that occur at one chain (intra) or between (inter) alginate chains.

3.3 SEM test

Figure 4 presents the results of visualization of the surface of the alginate membrane. The SEM image confirmed a microporous texture of the alginate surface membrane. This porous characteristics shows that the membrane is suitable for sustained release matrix formulation.

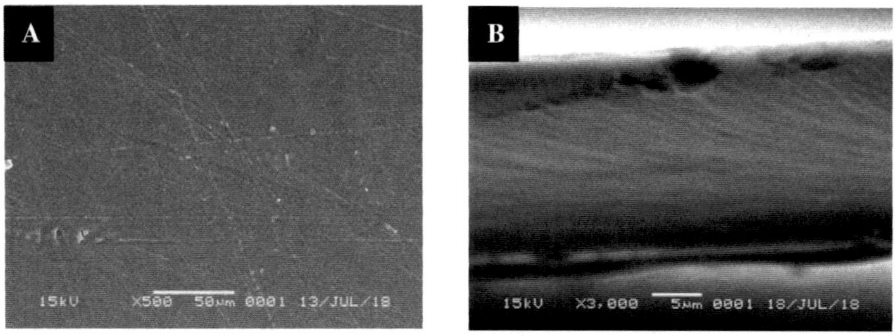

Figure 4. SEM test results: (A) Surface cross-section image with 500x magnification; (B) Surface cross-section image with 3000x magnification.

3.4 Mechanical properties test

From Table 1, the membrane with alginate concentration of 5% (w/v) has the largest modulus young than other concentration. The mechanical properties of membranes are getting better with increasing alginate concentrations. However, at the concentration higher than 5%, the modulus young is decreasing. This is because of its close structure causing the distance between the molecules in the membrane to be denser. At excessive alginate concentrations it may cause mechanical properties of the membranes to become fragile. The greater the value of Young's modulus, the better ability to prevent the damage caused by external forces, resulting in a stronger alginate membrane.

Tabel 1. Data of alginate membrane mechanical properties

Composition	% elongation	Young's Modulus (N/mm^2)
Alginate 3.3% (w/v)	9.03	8.84
Alginate 5% (w/v)	17.08	17.07
Alginate 10% (w/v)	6.69	7.36

4. Conclusion

The alginate membrane can be applied as matrix of sustained release drug formulation. The FTIR data showed that the alginate membrane has carbonyl group (C=O) and hydroxyl group (–OH). The SEM test results showed that the alginate membrane has a microporous surface. The results of mechanical tests showed the young modulus obtained in the membrane for variation of the concentration of alginate in water were 3.3% w/v, 5% w/v, and 10% w/v and the tensile strength of membranes indicated 8.84 MPa, 17.07 MPa, and 7.36 MPa respectively. The mechanical properties of membranes are better with increasing alginate concentration, but at larger concentration of alginate, the Young's modulus tend to decrease.

5. References:

[1] Majeti, N.V. and Kumar, R. 2000. *Functional Polymers*, 46, 1-27
[2] Shahidi, F. and Abuzaytoun, R. 2005. *Advances in Food and Nutrition Research*, 49, 93-135
[3] Honarkar, H. and Barikani, M. 2009. *Applications of biopolymers I: chitosan*, Published online: Springer-Verlag.

[4] Kanti, P., Srigowri, K., Madhuri, J., Smitha, B., and Sridhar, S. 2004. *Separation and Purification Technology*, 40, 259-266.
[5] Rokhati, N., Pramudono, B., Widiasa, I.N., dan Susanto, H. 2012. *Reaktor*, 14(2), 158-164
[6] Gong, Y., Han, G.T., Zhang, Y.M., Zhang, J.F., Jiang, W., Tao, X.W., and Gao, S.C. 2016. *J. Polym. Eng.*, 36(4), 363-370
[7] Kaban, J. 2006. *Jurnal Sains Kimia*, 10(1), 10-16.
[8] Duma, Y.E. and Cahyaningrum, S.E. 2012. *UNESA Journal of Chemistry*,1(2), 7-13
[9] Gombotz, W.R. and Wee, S. 1998. *Adv. Drug Deliv. Rev.*, 31, 267-285.
[10] Pasparakis, G. and Bouropoulos, N. 2006. *Int. J. Pharm.*, 323, 34-42.

Potential adsorption mechanisms of different bio-wastes to remove diazinon from aqueous solution

Sitha Sean[1], Quach An Binh[2], Dondej Tungtakanpoung[1], and Puangrat Kajityichyanukul[2]

1 Department of Environmental Engineering, Faculty of Engineering, Naresuan University, Phitsanulok 65000, Thailand
2 Centre of Excellence on Environmental Research and Innovation, Faculty of Engineering, Naresuan University, Phitsanulok, Thailand 65000
E-mail: kpuangrat@gmail.com

Abstract. This paper investigated the isotherms, kinetics, and mechanisms of three types of bio-waste derived from agro-waste, synthesized as potential adsorbents in diazinon extraction from aqueous solution. The adsorption capacity that differentiates each of the bio-wastes were discussed. Bio-wastes were synthesized for 4 h at a temperature of 600 °C. Bio-wastes were modified by hydrofluoric acid (HF) to enhance the specific area and porosity. The SEM, BET, FT-IR, and pHpzc were used to analyse the characteristics of the bio-wastes. Experimental results showed the different BET specific surface areas ranged from 67.42 to 402.43 $m^2.g^{-1}$ of each bio-waste. Meanwhile, SEM showed that bio-wastes possessed an intricate pore network comprising micropores and narrow mesopores. The pore volume disseminations in the size of the micropores (<2 nm) and narrow mesopores (2-20 nm) of bio-wastes in the range of 16.88-21.24% and 56.57-69.22%, respectively. Solution pH at 4 carried out the highest removal efficiency of diazinon. Langmuir isotherm model showed the best fit to the experimental adsorption data. The maximum adsorption capacities of bio-wastes ranged from 8.60-15.83 $mg.g^{-1}$. The pore filling and chemical adsorption participated in the adsorption mechanisms of diazinon and other bio-wastes.

1. Introduction

Pesticides are organic chemicals commonly used in agriculture to increase yield, and product quality [1]. Almost 40% of pesticides used internationally are organophosphorus compounds, which were utilized in huge amounts due to their affordability and high efficiency in pest removal [2]. Today, the contamination of surface and ground water by pesticides has become a significant problem [3]. Diazinon is an organophosphorus insecticide, which is commonly used against insects in agriculture. This pesticide can cause damage to humans, animals, and plants due to their consumption of water. Therefore, the presence of pesticides in water would be considered a serious issue that has to be solved. Recently, bio-wastes have become popular absorbents used in pesticide removal from the aqueous environment. With the high porosity, large surface area, ash contents, a mineral component, many functional groups, and great aromatic structure, bio-wastes were treated as an effective absorbent [4]. Many studies reported that bio-wastes are effective absorbents but have different levels of efficiency regarding conditions such as pyrolysis and the synthesis of conditions, porous structure, functional group, and mechanisms [5].

The objective of this study was to investigate the bio-waste materials obtained from three types of agro-waste, synthesized as potential adsorbents for diazinon removal from aqueous solutions and compare the adsorption capacity of each bio-waste. The adsorption was investigated in the batch adsorption system, including isotherm, kinetics, and mechanisms.

2. Material and methods

2.1 Chemicals and materials

The three types of bio-waste including coconut fibre bio-waste (CF-BW), corn cob bio-waste (CC-BW), and bagasse bio-waste (BG-BW) were collected in Northern Thailand. The diazinon (60% w/v) was provided by the SIMS AGROW CHEME CO.,LTD. HF (48.0%) from Loba Chemie Pvt. Ltd, India and deionized water with a resistivity of 18 MΩ.cm.

2.2 Bio-waste preparation

Bio-wastes were sliced into the approximate size (cm^{-1}) of 1 x 1 and dried in the oven at 105 °C for a few hours until getting the constant weight. The bio-wastes were burned in a furnace (Nabertherm, Germany) with the temperature slowly increased by 3 °C per minute. The pyrolysis process was conducted under oxygen-limited conditions at 600 °C for 4 hours. The bio-wastes were washed with HF acid by following the method of Minori Uchimiya group [6].

2.3 Bio-wastes characterization

The surface morphology and specific surface area of bio-wastes were analysed by the Scanning Electron Microscope (SEM) and the Brunauer–Emmett–Teller methods (BET), respectively. The FT-IR method was applied to define the functional groups of adsorbents. Mass titration method was utilized to define the point zero charge (pHpzc) of bio-wastes [7].

2.4. Adsorption procedure

The volume of the pesticides and bio-wastes of 1.5 $g.L^{-1}$ was added into 250 ml Erlenmeyer flasks and shaken by MS Orbital Shaker (MS-NOR-30/MS-NOR-3001) for equilibrium time at 120 rpm in dark conditions. A UV–Vis spectrometer (Genesys 10S UV-Vis spectrophotometer, Thermo Scientific, USA) was used to analyse the residual concentration of diazinon by using the wavelength of 247 nm. Whatman filter paper 42 with the retention of 2.5 μm was used to separate the bio-wastes from the diazinon solution. Furthermore, the effect of pH of the solution was prepared in the range from 2.0 to 10.0. HCl and NaOH solutions were used to adjust the pH of the pesticide solutions. The batch adsorption method was applied to the diazinon adsorption experiment.

3. Results and discussion

3.1 Characterization of the bio-wastes

The SEM of the bio-wastes are shown in Figure 1. The pore size of all types of bio-waste had tiny pores and no mineral on the surface areas. The size of the surface morphology of the bio-wastes is the crucial element in the adsorption process.

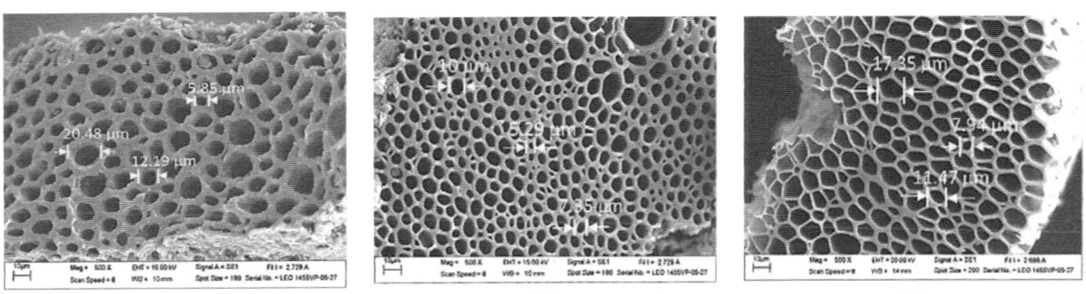

(a) (b) (c)

Figure 1. SEM pictures of bio-wastes (a,b,c): CC-BW, BG-BW, CF-BW, respectively.

The specific surface area (BET), total pore volume (TPV), and pore volume distributions of bio-wastes were displayed in Table 1. Results show that the specific surface area of CF-BW is higher than CC-BW and BG-BW in the ordering value 402.43 $m^2.g^{-1}$, 292.92 $m^2.g^{-1}$, and 67.42 $m^2.g^{-1}$, respectively. Noticeably, the BET is the main criteria to determine the adsorption capacity of bio-wastes. If the large specific area exists on the surface of the bio-wastes, it has more interfaces for pollution adsorption [8]. Likewise, the total pore volume of CF-BW is also higher than other bio-wastes by distribution into four sizes including, micropores (<2 nm), narrow mesopores (2-20 nm), mesopores (21-50 nm), and macropores (>50 nm). The majority pore volumes of bio-wastes were narrow mesopores and micropores. The TPV is the main factor for the high adsorption of bio-wastes. The pore size of bio-wastes is very important in achieving the uptake of adsorbate.

Table 1. The BET (SSA), TPV, and the pore volume disseminations in pore size of bio-wastes.

Property	Unit	BG-BW	CC-BW	CF-BW
BET suface area	$m^2.g^{-1}$	67.42	292.92	402.43
TPV	$cm^3.g^{-1}$	0.029	0.117	0.151
Micropores (< 2 nm)	%	16.88	21.24	17.40
Narrow mesopores (2-20 nm)	%	59.29	56.57	69.22
Mesopores (21-50 nm)	%	15.75	10.39	6.10
Macropores (> 50 nm)	%	8.07	11.81	7.28

The results of the surface functional groups of adsorbents were demonstrated by FT-IR analysis. The FT-IR spectra of the bio-wastes are exhibited in Figure 2.

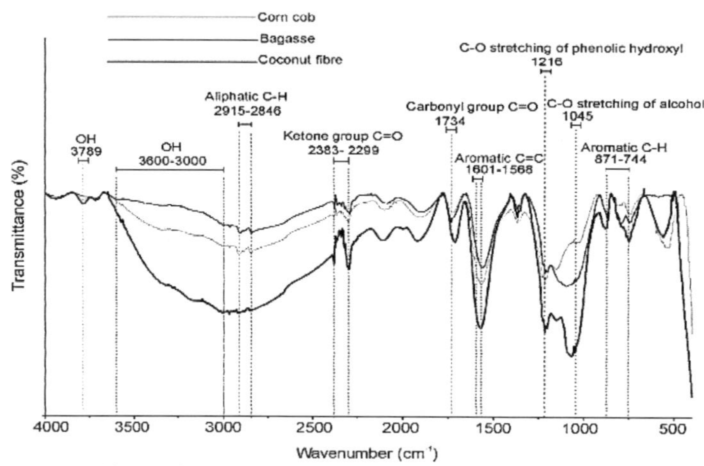

Figure 2. FT-IR spectrum analysis of bio-wastes.

On the surfaces of the bio-wastes are the eight main functional groups in a different spectrum. The band 3000-3600 cm^{-1} and peak at 3789 cm^{-1} were hydroxyl groups (O-H) stretching vibrations of carboxylic acid, alcohol, and phenol [9]. Moreover, the hydroxyl groups (O-H) are presented with significant H-bonding interaction. Two types of groups C-H such as aromatic C-H and Aliphatic group were found in this adsorption. Aromatic C-H bands were around 744-871 cm^{-1} and Aliphatic C-H bands were 2846-2915 cm^{-1}. The bands at 1216 cm^{-1} and 1217 cm^{-1} represented the aromatic C-O stretching of phenolic hydroxyl [10]. The bands at 1045 cm^{-1}, 1070 cm^{-1}, and 1099 cm^{-1} were recognized to the aromatic C-O stretching of alcohol [11]. The peaks around 1734 cm^{-1} and 1715 cm^{-1} were assigned to the carbonyl group C=O and could show ketone, carboxylic acid, and ester. These groups can cause H-bonding interaction [12]. The bands at 2299-2383 cm^{-1} were ascribed to the ketone group C=O [13]. The bands about 1568-1601 cm^{-1} were ascribed to the aromatic C=C [11]. According to FT-IR spectrum analysis, the major functional groups of the adsorbents include ketone, carbonyl, and aromatic organic molecules which are unsaturated hydrocarbons. The hydroxyl group is also the main molecule on the adsorbents. All these functional groups can cause H-bonding interaction between adsorbents and pollutants [12]. Furthermore, the aromatic group of the adsorbents has the π-π EDA interaction with the pollutant [12] and alkyl group has hydrophobic properties [14].

3.2. Effect of the pH on diazinon adsorption

The pH of solution influenced the diazinon removal was indicated in Figure 3. From the result, it can be observed that the efficiency of diazinon removal increased when pH values increased to pH 4 but decreased gradually when the pH values are higher than pH 4. However, the electrostatic interaction was engendered by the surface of bio-wastes and diazinon, depending on pH changes in the solution. The percentage of the highest removal of diazinon was succeeded at the pH 4.0, which was not performed by the electrostatic attraction. As a reason, at pH 4, the diazinon was formed as negatively charged due to the pH of the solution was higher than the pKa (2.6). The surface charges of the bio-wastes were negatively charged when the pH of the solution was adjusted higher than the pHpzc (0.7) value. The adsorption of diazinon by bio-waste occurred without electrostatic attraction. However, the hydrogen bonding, hydrophobic, pore-filling, and π-π EDA interactions can act as factors in this adsorption.

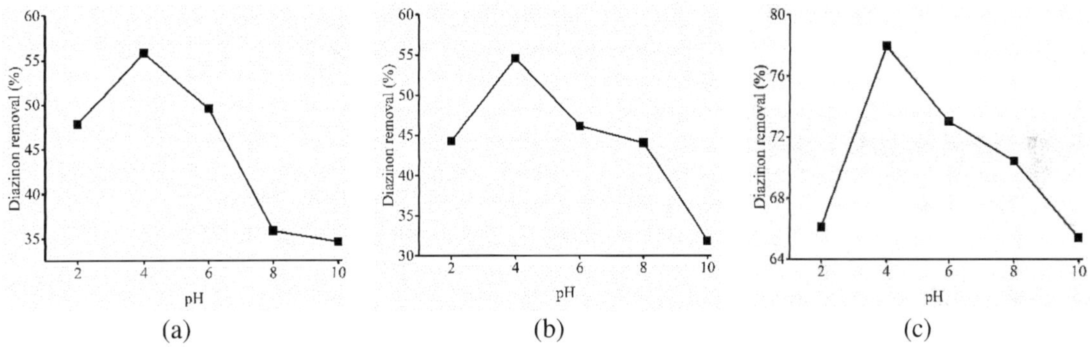

Figure 3. The influence of pH on the efficiency of diazinon removal by bio-wastes (a) CC-BW, (b) BG-BW, and (c) CF-BW.

3.3. The isotherm adsorption models

The adsorption isotherms including Langmuir and Freundlich models of diazinon and bio-wastes were shown in Table 2. The results showed that the Langmuir model of bio-waste adsorption was more appropriate than the Freundlich model due to the R_2 of the Langmuir model were higher than the R_2 Freundlich model. These factors demonstrated that the monolayer adsorption of diazinon occurred on the surface of bio-wastes. Regarding R_L values in the adsorptions were lower than 1, illustrated the form of the isotherm is favourable for adsorption.

Table 2. The isotherm of the adsorptions.

Models	Parameters	Bio-wastes		
		CC-BW	BG-BW	CF-BW
Langmuir	Q_0 (mg.g^{-1})	17.60	12.00	24.60
	K_L (L.mg^{-1})	0.050	0.210	0.190
	R_L	0.434	0.163	0.174
	R_2	0.982	0.928	0.993
Freundlich	K_F (mg.g^{-1}) (mg.L^{-1})$^{-n}$	1.231	3.045	4.650
	1/n	0.641	0.458	0.581
	R_2	0.956	0.917	0.966

3.4. Effects of the initial diazinon concentration on to the adsorption process

The rate of the adsorption process and initial diazinon concentration had a parallel correlation. The result in Figure 4 showed that increasing the diazinon concentration brought about the increase in adsorption efficiency. The driving force of the concentration gradient was strengthened when the initial diazinon concentration was increased, generating a higher adsorption capacity [15]. Moreover, the adsorption capacity of CF-BW was higher than another two of bio-wastes.

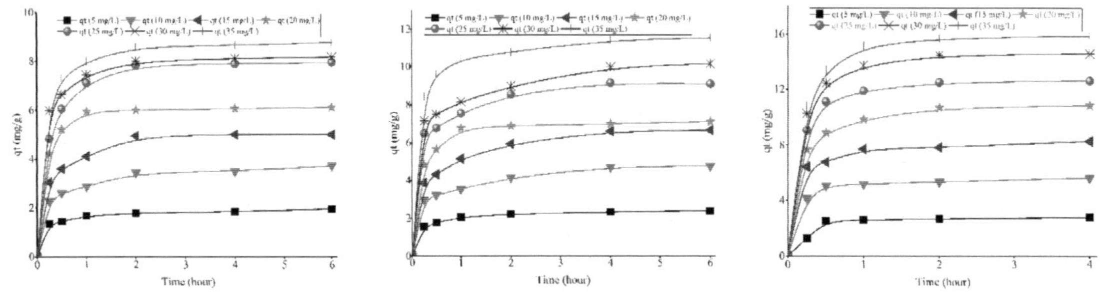

CC-BW BG-BW CF-BW

Figure 4. The adsorption capacities of diazinon by bio-wastes (The dosage of adsorbents 1.5 g.L^{-1}, pH = 4).

3.5. The adsorption mechanism of bio-wastes and diazinon

The hydrogen bonding, electrostatic, hydrophobic, pore-filling, and π-π EDA interactions were the adsorption mechanism between bio-wastes and organic pollutants [16]. In this study, the SSA and the TPV results of bio-wastes before and after diazinon adsorption were different. This validated that the pore-filling mechanism and chemical interaction occurred between adsorbents and diazinon adsorption. According to the pore size data from Table 1, the CC-BW, BG-BW, and CF-BW had a high distribution of micropores (<2 nm) and narrow mesopores (2-20 nm), in the range between 21.24 and 56.57 %, 16.88 and 59.29%, and 17.40 and 69.22%, respectively. The dimension of the molecular structure of diazinon was 1.08 x 0.58 x 1.04 nm (0.65 nm^3) analysed by using the ChemBio3D software. The data of the bio-wastes pore size were quite close to the structural size of diazinon. Hence, the diazinon can be adsorbed into micropores and narrow mesopores of the bio-wastes. The pore-filling mechanism happened when the pore size of adsorbent is close to the molecular structure size of adsorbate [17].

The proposed chemical interaction of diazinon and bio-wastes adsorption contain H-bonding, hydrophobic, π-π EDA interactions regarding the results of FT-IR (Figure 5).

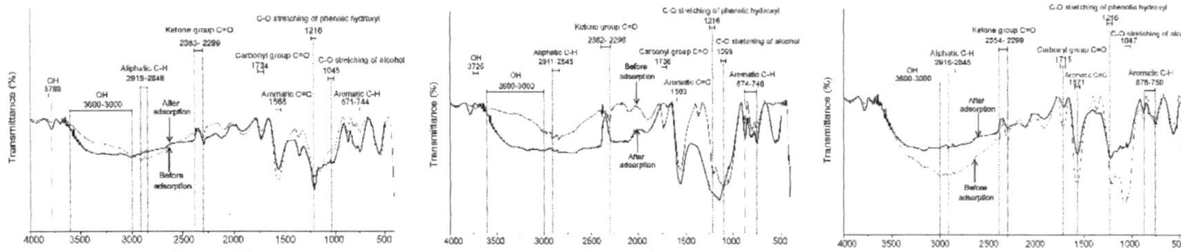

CC-BW BG-BW CF-BW

Figure 5. The FT-IR analysis of bio-wastes before and after diazinon adsorption.

Figure 5 displays the formation of H-bonding and the shift of FT-IR results after diazinon adsorption. The peak and peak area were assigned to the C-O stretching of alcohol (1045 cm^{-1}) and phenolic hydroxyl (1216 cm^{-1}), C=O of carbonyl group (1734 cm^{-1}), C=O of ketone group (2383-2299 cm^{-1}), O-H of alcohol, phenol, and carboxyl (3600-3000 cm^{-1}) which were changed as shown in Figure 5. All of the peaks suggest that O and H atoms in these functional groups were involved in H-bonding interactions.

The π-π EDA interactions of the bio-wastes after diazinon adsorption is shown in Figure 5. The changing of the peaks showed that the aromatic C=C which was related to the π-π EDA interactions. The hydroxyl group of bio-wastes was an electron providing group, this leads to the aromatic rings of bio-wastes to be high in electron density (π^-). The halogen group had the electron-withdrawing inductive, which caused the nucleus to carry a positive charge (π^+). So, π-π EDA interactions were formed between diazinon and the aromatic hydrocarbon compound of the bio-wastes. Several functional groups of bio-wastes including, alcohol, phenol, and carboxylic acid groups, showed the H-bonding interaction. Whereas, the alkyl group of bio-wastes presented the hydrophobic interaction and the hydroxyl group of bio-wastes presented the π-π EDA interaction. Therefore, the H-bonding interaction was the major chemical interaction. To conclude, the adsorption mechanism between the bio-wastes and diazinon consisted of the pore-filling sorption mechanism and the chemical interaction. H-bonding was the primary interaction that played an important role in chemical adsorption.

4. Conclusion

This study examined the adsorption mechanism of bio-wastes for diazinon removal. pH 4 was the optimum pH of the solution that engendered the highest removal of diazinon. The adsorption isotherm was followed by the Langmuir isotherm model. When the initial diazinon concentration was increased, this enhanced the adsorption capacity of bio-wastes. The pore-filling adsorption and the chemical interaction were the main factors in the adsorption mechanism. The chemical interaction included H-bonding, hydrophobic, and the π-π EDA interactions; however, the main contributor was H-bonding. Overall, bio-wastes were very useful materials and had a high potential to contribute to the treatment of environmental pollutants.

5. References

[1] G. P. Satish, D. M. Ashokrao, and S. K. Arun, "Microbial degradation of pesticide: A review," *African Journal of* Microbiology *Research,* **vol. 11**, pp. 992-1012, 2017.

[2] M. H. Dehghani, S. Kamalian, M. Shayeghi, M. Yousefi, Z. Heidarinejad, S. Agarwal, *et al.*, "High-performance removal of diazinon pesticide from water using multi-walled carbon nanotubes," *Microchemical Journal,* **vol. 145**, pp. 486-491, 2019.

[3] K. Kuśmierek, M. Szala, and A. Świątkowski, "Adsorption of 2,4-dichlorophenol and 2,4-dichlorophenoxyacetic acid from aqueous solutions on carbonaceous materials obtained by combustion synthesis," *Journal of the Taiwan Institute of Chemical Engineers,* **vol. 63**, pp. 371-378, 2016.

[4] M. S. Khorram, D. Lin, Q. Zhang, Y. Zheng, H. Fang, and Y. Yu, "Effects of aging process on adsorption-desorption and bioavailability of fomesafen in an agricultural soil amended with rice hull biochar," *J Environ Sci (China),* **vol. 56**, pp. 180-191, Jun 2017.

[5] M. Inyang and E. Dickenson, "The potential role of biochar in the removal of organic and microbial contaminants from potable and reuse water: A review," *Chemosphere,* **vol. 134**, pp.232-40, Sep 2015.

[6] M. Uchimiya, L. H. Wartelle, and V. M. Boddu, "Sorption of triazine and organophosphorus pesticides on soil and biochar," *J Agric Food Chem,* **vol. 60**, pp. 2989-97, Mar 28 2012.

[7] J. S. Noh and J. A. Schwarz, "Effect of HNO3 treatment on the surface acidity of activated carbons," *Carbon,* **vol. 28**, pp. 675-682, 1990/01/01/ 1990.

[8] G. Tan, W. Sun, Y. Xu, H. Wang, and N. Xu, "Sorption of mercury (II) and atrazine by biochar, modified biochars and biochar based activated carbon in aqueous solution," *Bioresource Technology,* **vol. 211**, pp. 727-735, 2016/07/01/ 2016.

[9] H. N. Tran, Y.-F. Wang, S.-J. You, and H.-P. Chao, "Insights into the mechanism of cationic dye adsorption on activated charcoal: The importance of π–π interactions," *Process Safety and Environmental Protection,* **vol. 107**, pp. 168-180, 2017/04/01/ 2017.

[10] Y. Chun, G. Sheng, C. T. Chiou, and B. Xing, "Compositions and Sorptive Properties of Crop

Residue-Derived Chars," *Environmental Science & Technology,* **vol. 38**, pp. 4649-4655, 2004/09/01 2004.

[11] M. Saffari, N. Karimian, A. Ronaghi, J. Yasrebi, and R. Ghasemi-Fasaei, *Stabilization of nickel in a contaminated calcareous soil amended with low-cost amendments* **vol. 15**, 2015.

[12] N. Liu, A. B. Charrua, C.-H. Weng, X. Yuan, and F. Ding, "Characterization of biochars derived
from agriculture wastes and their adsorptive removal of atrazine from aqueous solution: A comparative study," *Bioresource Technology,* **vol. 198**, pp. 55-62, 2015/12/01/ 2015.

[13] K. Nuithitikul, S. Srikhun, and S. Hirunpraditkoon, "Kinetics and equilibrium adsorption of Basic Green 4 dye on activated carbon derived from durian peel: Effects of pyrolysis and post-treatment conditions," *Journal of the Taiwan Institute of Chemical Engineers,* **vol. 41**, pp. 591-598, 2010/09/01/ 2010.

[14] T. J. Kinney, C. A. Masiello, B. Dugan, W. C. Hockaday, M. R. Dean, K. Zygourakis, *et al.*, "Hydrologic properties of biochars produced at different temperatures," *Biomass and Bioenergy,* **vol. 41**, pp. 34-43, 2012/06/01/ 2012.

[15] C.-H. Weng, Y.-T. Lin, and T.-W. Tzeng, "Removal of methylene blue from aqueous solution by adsorption onto pineapple leaf powder," *Journal of Hazardous Materials,* **vol. 170**, pp. 417-424, 2009/10/15/ 2009.

[16] X. Tan, Y. Liu, G. Zeng, X. Wang, X. Hu, Y. Gu, *et al.*, "Application of biochar for the removal of pollutants from aqueous solutions," *Chemosphere,* **vol. 125**, pp. 70-85, 2015/04/01/ 2015.

[17] F. Lian, G. Cui, Z. Liu, L. Duo, G. Zhang, and B. Xing, "One-step synthesis of a novel N-doped microporous biochar derived from crop straws with high dye adsorption capacity," *Journal of Environmental Management,* **vol. 176**, pp. 61-68, 2016/07/01/ 2016.

Acknowledgments
The author, Sitha Sean received scholarship for his Master Degree from the Royal Scholarship of Naresuan University. This work is supported by the Thailand Research Fund (Grant No. BRG 6180009) and National Research Council, Thailand (NRCT) through Naresuan University (Grant No. R2562B022).

Analysis of a cylindrical cyclone separator based on the RMS data

Z W Gao[1,2], J Wang[1,2], J Y Wang[1,2], Y Mao[1], Y D Wei[1,2]

1 State Key Laboratory of Heavy Oil Processing, China University of Petroleum, Beijing 102249, China.
2 Beijing Key Laboratory of Process Fluid Filtration and Separation, Beijing 102249, China
E-mail: wangjuan@cup.edu.cn

Abstract. This study presents a flow analysis on an atypical cylindrical cyclone separator with a large ratio of length to diameter. The Phase Doppler Particle Analyzer (PDPA) is used to measure the gas flow, and the numerical simulation is carried out by using the Reynolds Average Navier-Stokes (RANS) equation, the Reynolds stress model (RSM). Since the model is validated by good agreement between the numerical results and experimental data, and then the tangential velocity is analysed from the transient flow field and dynamic performance. The results showed that the tangential velocity of the transient flow field existed a non-axisymmetric phenomenon in the cross section, which was mainly presented as asymmetric distribution of the contours. The zero location of tangential velocity value was not coincided with the geometric center, where the tangential velocity was larger near the zero location. In addition, the dynamic character of tangential velocity was high-speed pulsating, with quasi-periodic characteristic. Meanwhile, a parameter of RMS was used to identify the degree of turbulence velocity deviating from mean velocity. The RMS data showed that the flow instability gradually increased with the fluid flowing into the cyclone separator from the inlet section, until reached a peak value at the position about 2.0 times the length to diameter. And then the flow instability gradually decreased, with the fluid energy dissipated.

1. Introduction

Cyclone separator is an important equipment for gas-solid separation process. Although there is no moving parts inside, the internal flow field is a complex turbulent flow. The swirling flow is instability, where the flow frequency changes with time [1]. But the distribution of flow field determines the separation performance. The instability of the swirling flow not only affects the distribution of the flow field, but also causes noise and the pressure fluctuation of the shell to form the vibration [2]. Therefore, the research of flow instability in cyclone separator has a far-reaching significance for engineering.

For the flow instability of in cyclone separator, many scholars have experimentally [3-5] and numerically simulated them [6-8]. However, most scholars analyzed the cyclone separator, still focused on the transient flow field, while neglecting the research of dynamic change of swirling fluid and how to identify the degree of the flow instability.

Aiming to understand the flow behaviour, this work will conduct a numerical investigation of the hydrodynamics in the cyclone separator with a large ratio of length to diameter. The RSM model will be used to capture the flow. The key is focus on the understanding of the relationship between the velocity fluctuation and the flow instability. This work can provide further insight into the design and optimization, which is highly significant.

IOP Conf. Series: Materials Science and Engineering **617** (2019) 012013 doi:10.1088/1757-899X/617/1/012013

2. System description and numerical methods

2.1. Configuration of the cyclone separator

The schematic diagram of the cyclone separator was shown in Figure 1 and the parameters characterizing the structure were given in Table 1. The cylinder of the cyclone separator was 140 mm in diameter. The cyclone with a large ratio of length to diameter, which doesn't have a section of the cone, could well reflect the flow of swirling fluid. During simulation analysis, six monitoring surfaces were considered, including *-z/D*=0.71, 2.14, 3.57, 6.43, 9.29 and 12.14 respectively. The horizontal plane across the bottom of the entrance was set as the reference plane, and the positive direction was upward. While the geometric center of the vortex finder was set as the origin.

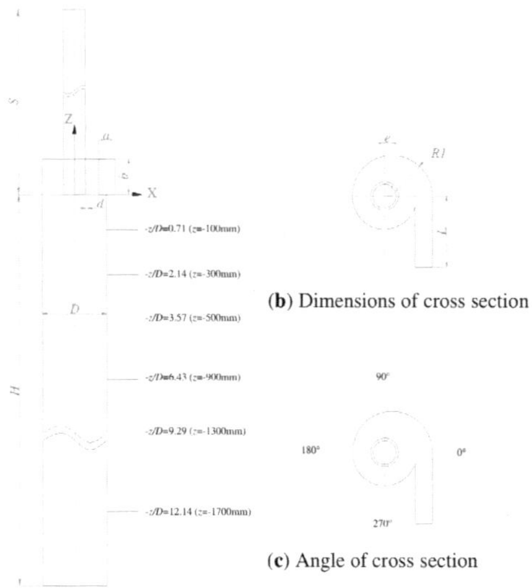

(**b**) Dimensions of cross section

(**c**) Angle of cross section

(**a**) Dimensions of vertical section

Figure 1. Dimensions of the cyclone separator.

The schematic diagram of the cyclone separator was shown in Figure 1 and the parameters characterizing the structure were given in Table 1. The cylinder of the cyclone separator was 140 mm in diameter. The cyclone with a large ratio of length to diameter, which doesn't have a section of the cone, could well reflect the flow of swirling fluid. During simulation analysis, six monitoring surface was considered, including *-z/D*=0.71, 2.14, 3.57, 6.43, 9.29 and 12.14 respectively. The horizontal plane across the bottom of the entrance was set as the reference plane, and the positive direction was upward. While the geometric center of the vortex finder was set as the origin.

Table 1. Geometry dimensions of the cyclone separator

Symbol	**Dimensions/mm**
D	140
H	1919
De	50
a	36
b	76
S	1000
L	150

e	16
d	5
R_1	86

2.2. Governing equations

Previous studies [9] have determined the Reynolds stress model (RSM) as having the greatest potential for predicting the swirling flow in cyclone separator. For incompressible fluid flow in cyclone separators, the governing equations can be expressed as:

$$\frac{\partial v_i}{\partial x_i} = 0 \tag{1}$$

$$\rho v_j \frac{\partial v_i}{\partial x_j} = -\frac{\partial p}{\partial x_i} + \frac{\partial}{\partial x_j}\left(\mu\left(\frac{\partial v_i}{\partial x_j} + \frac{\partial v_j}{\partial x_i}\right)\right) + \rho g_i - \frac{\partial}{\partial x_j}\left(\rho \overline{v_i' v_j'}\right) \tag{2}$$

where p, ρ and μ are pressure, density and dynamic viscosity, respectively. v and v' are the mean and fluctuating velocity. The Reynolds stress term, $\rho\overline{v_i' v_j'}$, represents the effect of turbulent fluctuations on the fluid flow and is modelled by using RSM in this study. The transport equation of Reynolds stress is given below:

$$\frac{\partial}{\partial x_k}\rho v_k \overline{v_i' v_j'} = P_{ij} + \phi_{ij} + \frac{\partial}{\partial x_k}\left(\left(\mu + \frac{\mu_t}{\sigma_k}\right)\frac{\partial}{\partial x_k}\left(\overline{v_i' v_j'}\right)\right) - \frac{2}{3}\rho\varepsilon\delta_{ij} \tag{3}$$

where P_{ij} is the stress production term given by

$$P_{ij} = -\rho\left(\overline{v_i' v_k'}\frac{\partial v_j}{\partial x_k} + \overline{v_j' v_k'}\frac{\partial v_i}{\partial x_k}\right) \tag{4}$$

and ϕ_{ij} is the pressure-strain term, which is modeled by

$$\phi_{ij} = -C_1\rho\frac{\varepsilon}{k}\left(\overline{v_i' v_j'} - \frac{2}{3}k\delta_{ij}\right) - C_2\left(P_{ij} - \frac{1}{3}P_{kk}\delta_{ij}\right) \tag{5}$$

In addition, the transport equation for turbulence dissipation rate ε is expressed as

$$\frac{\partial}{\partial x_i}\left(\rho v_i \varepsilon\right) = \frac{\partial}{\partial x_j}\left(\left(\mu + \frac{\mu_t}{\sigma_\varepsilon}\right)\frac{\partial \varepsilon}{\partial x_j}\right) + \frac{\varepsilon}{k}\left(\frac{1}{2}C_{\varepsilon 1}P_{ii} - C_{\varepsilon 2}\rho\varepsilon\right) \tag{6}$$

The following constants of RSM are used, C_1=1.8, C_2=0.6, $C_{\varepsilon 1}$=1.44, $C_{\varepsilon 2}$=1.92, σ_k=1, σ_ε=1.3.

3. Numerical simulation

The ANSYS ICEM software was used to mesh and ensure all the structured grid for hexahedron grid. The whole computational domain was divided into 234610 elements. The simulation result of this grid domain was examined to be grid independent with three grids of different densities.

Following actual operating conditions, a constant inlet velocity at the inlet boundary surface and a constant pressure at the outlet boundary surface were used. In addition, the fluid flow was modelled using the standard wall function and no-slip boundary condition near the wall.

The governing equations mentioned above were solved by a finite volume method. These equations were discretized by the QUICK scheme, and solved by the commercial CFD code FLUENT 16.0. The SIMPLE algorithm was implemented to deal with the velocity-pressure coupling. The numerical solution was considered converged when the scaled residuals of the continuity equation was below 1×10^{-4} s.

4. Results and discussion

4.1. Model validation

For cyclone separator, tangential velocity and axial velocity plays a critical role in the separation process. To validate the model of this study, the numerical results of the two velocities were compared with experimental data. As shown in Figure 2, both of them were in good agreement, which indicated that the model of this study has a good prediction accuracy for the flow field distribution in cyclone separator.

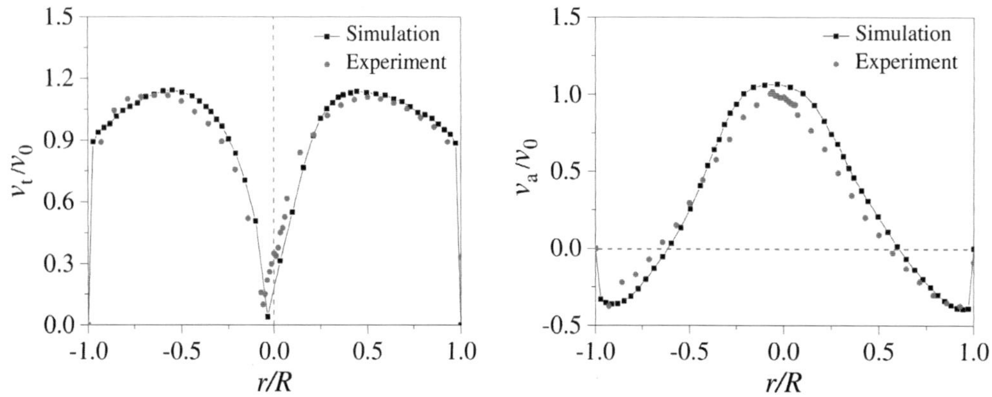

(a) Tangential velocity (b) Axial velocity

Figure 2. Comparison of simulation result and experimental data.

4.2. Transient flow field analysis

Tangential velocity dominates the flow behavior in the cyclone separator, so tangential velocity was analyzed. Figure 3 shows the tangential velocity distribution. From the longitudinal section, the tangential velocity was divided into the inner and outer double-layer structure at different axial positions. Each section had a rotation center, the center and the geometric center was not completely coincide. In the process of the fluid moving, the position of the rotation center was constantly changing. And then the procession vortex core (PVC) phenomenon was formed. At this time, the movement of fluid was not periodicity, but a complex movement with a strong unsteady characteristic.

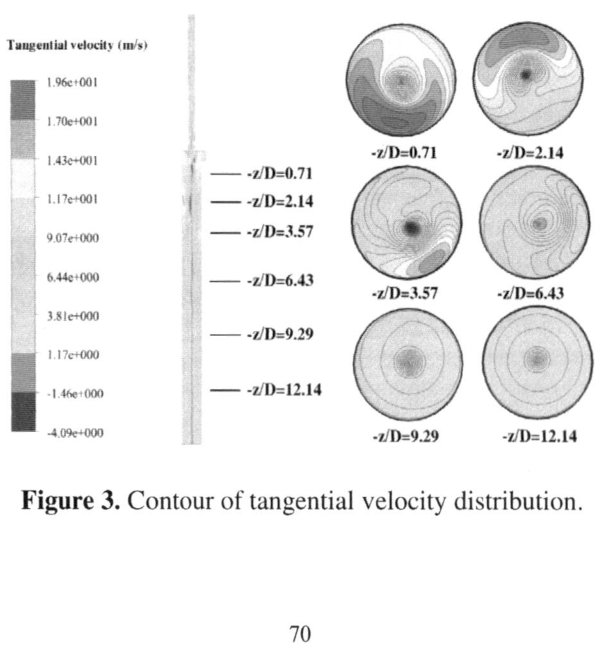

Figure 3. Contour of tangential velocity distribution.

4.3. Dynamic characteristics analysis

The gas-solid separation process in the cyclone is a dynamic continuous process. Figure 4 shows the fluctuation curve of the instantaneous tangential velocity pulsation value. In the upper part of the cylinder, the fluctuation range of tangential velocity was -8~8 m/s, due to the flow instability. With the axial downward, the fluid energy gradually dissipated, tangential velocity pulsation range gradually reduced.

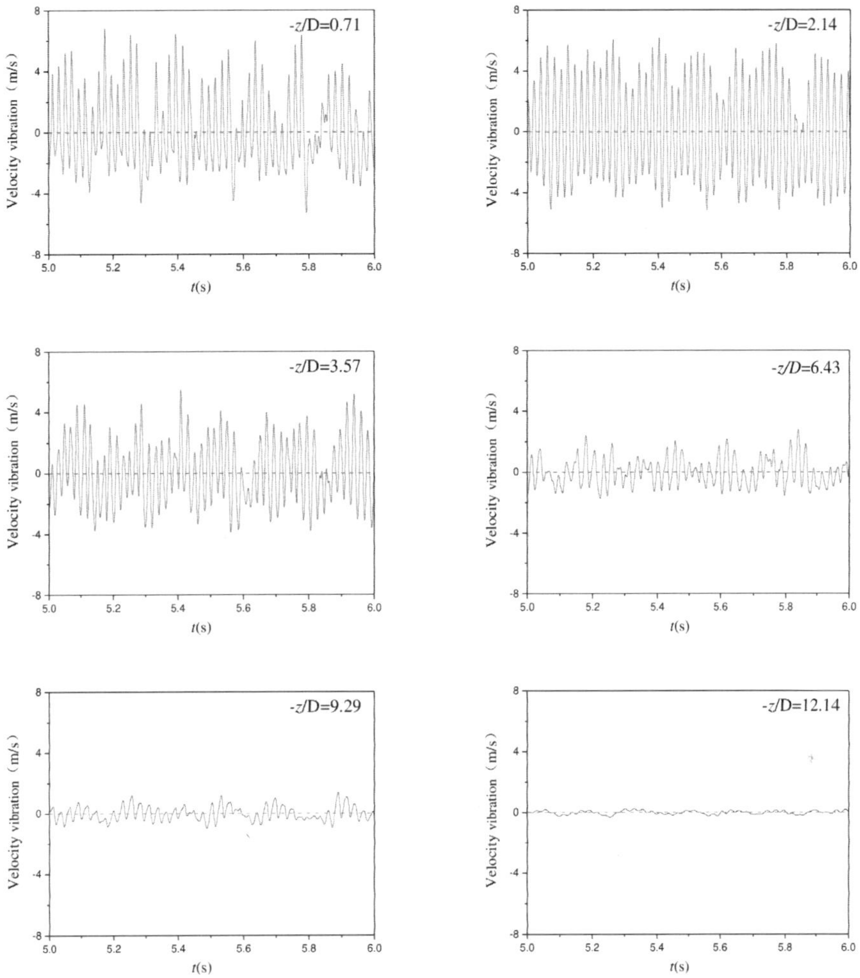

Figure 4. Curve of tangential velocity fluctuation, r/R=0.14.

4.4. Swirling RMS analysis

Turbulence pulsating motion was irregularly random in time series. In order to study the flow instability in the cyclone, the time-averaged velocity of each section was recorded with time, and the deviation of the sample and the mean could be obtained by analyzing the standard deviation. The RMS value of the turbulence pulsating velocity of the measured point could be expressed by the following formula, which reflects the degree of velocity fluctuation.

$$\mathrm{RMS} = \sqrt{\frac{\sum_{i=1}^{n}\left(v_i - \bar{v}\right)^2}{N}} = \sqrt{\frac{\sum_{i=1}^{n} v_i'^2}{N}} \quad (7)$$

IOP Conf. Series: Materials Science and Engineering **617** (2019) 012013 doi:10.1088/1757-899X/617/1/012013

Where N was the number of measurements, v_i was the instantaneous velocity, $\bar{v}$ was the mean velocity, and v_i' was the pulsation velocity of the measurement sample.

Before using the RMS analysis of the cyclone instability of the cyclone, it was necessary to verify the sampling independence. As shown in Figure 5, the results showed that the RMS distributions were basically the same in two sampling cases.

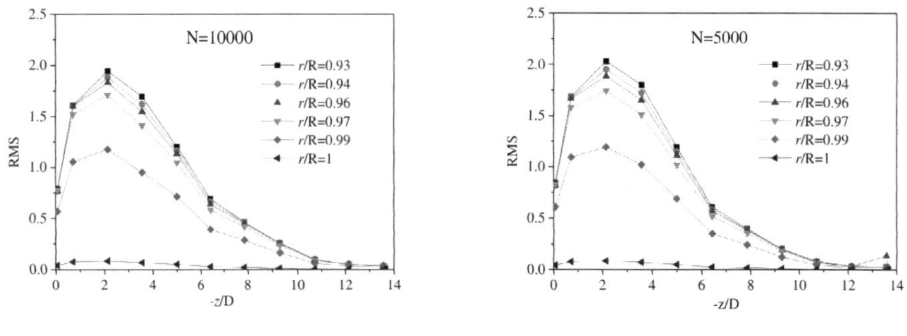

Figure 5. The curve of the sampling independent verification.

Figure 6 shows the distribution of RMS in cyclone separator. It could be found that the value of RMS increased from the inlet section, until reached to a peak. Which indicated that the movement of fluid from the inlet section into the cyclone separator was unsteady, and the flow instability reach a maximum in a certain position.

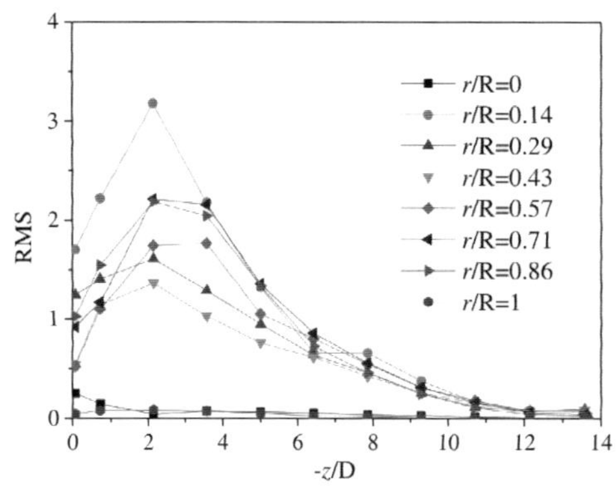

Figure 6. The distribution curve of RMS in cyclone separator.

5. Conclusions

By numerical simulation and experiments, this paper aimed to study the flow instability character in cyclone separator. Moreover, a cylindrical cyclone with a lager ratio of length to diameter was analyzed to describe the performance. The main conclusion could be summarized as follows:

(1) The tangential velocity zero value does not coincide with the geometric center. The movement of fluid was not periodicity, but a complex movement with a strong unsteady characteristics.

(2) In the upper space of the separation space, the instability of the moving fluid is larger, the turbulence pulsation is stronger and the velocity fluctuation range is larger. With the axial downward, the fluid energy gradually dissipated, the speed of pulsation gradually reduced.

(3) Using the RMS data to analyze the degree of velocity fluctuation. The movement of fluid from the inlet section into the cyclone separator was unsteady, and the flow instability reach a maximum in a certain position.

6. *References*

[1] Gao Z.W., Wang J., Wang J.Y., Mao Y. and Wei Y.D. (2017), "Vortex analysis of flow field of cyclone separator with single volute inlet structure", *CIESC Journal*, 68(8), pp. 3006-3013.
[2] Song J.F., Wei Y.D., Sun G.G. and Chen J.Y. (2017), "Experimental and CFD study of particle deposition on the outer surface of vortex finder of a cyclone separator", *Chemical Engineering Journal*, 309, pp. 249-262.
[3] Hoekstra A.J., Derksen J.J., and Vandenakker H.E.A (1999), "An experimental and numerical study of turbulent swirling flow in gas cyclones", *Chemical Engineering Science*, 54, pp. 2055-2065.
[4] Derksen J.J. and Vandenakker H.E.A. (2000), "Simulation of vortex core precession in a reverse-flow cyclone", *AIChE Journal*, 46(7), pp. 1317-1331.
[5] Hoffermann A.C., Peng W. and Dries H. (2006), "Effect of pressure recovery vanes on the performance of a swirl tube with emphasis on the flow pattern and separation efficiency", *Energy Fuel*, 20(4), pp. 1691-1697.
[6] Chu K.W., Wang B., Xu D.L., Chen Y.X. and Yu A.B. (2011), "CFD-DEM simulation of the gas-solid flow in a cyclone separator", *Chemical Engineering Science*, 66(5), pp. 834-847.
[7] Gao J.S., Chang J., Xu C.M., Lan X.Y. and Yang, Y. (2008), "CFD simulation of gas solid flow in FCC strippers", *Chemical Engineering Science*, 63(7), pp. 1827-1841.
[8] Elsayed K. (2015), "Optimization of the cyclone separator geometry for minimum pressure drop using *Co*-Kriging", *Powder Technology*, 269, pp. 409-424.
[9] Gao Z.W., Wang J., Wang J.Y., Mao Y., and Wei Y.D. (2019), "Analysis of the effect of vortex on the flow field of a cylindrical cyclone separator", *Separation and Purification Technology*, 211, pp. 438-447.

IOP Conf. Series: Materials Science and Engineering **617** (2019) 012014 doi:10.1088/1757-899X/617/1/012014

Enhancement of color quality scale and luminous flux of white LEDs with remote phosphor geometries

Nguyen Thi Phuong Loan[1], Duc Lam Pham[2], Nguyen Doan Quoc Anh[3*]

[1]Posts and Telecommunications Institute of Technology, Ho Chi Minh City, Vietnam
[2]Faculty of Mechanical, Electrical, Electronic and Automotive Engineering, Nguyen Tat Thanh University, Ho Chi Minh City, Vietnam
[3]Power System Optimization Research Group, Faculty of Electrical and Electronics Engineering, Ton Duc Thang University, Ho Chi Minh City, Vietnam
E-mail: nguyendoanquocanh@tdtu.edu.vn

Abstract. In comparison with the conformal phosphor or in-cup phosphor structure, remote phosphor structure has a superior luminous flux whereas its color quality is inferior to the others'. Due to this drawback, it is essential to conduct experiments to find the solution for the improvement of the color quality of WLEDs using remote phosphor structure. In this study, a dual-layer remote phosphor structure that is capable of enhancing the color rendering index (CRI) and color quality (CQS) for WLEDs was proposed. Three kinds of WLEDs with similar structures but different color temperatures varying at 5600 K, 7000 K, and 8500 K are used in this article. The mission of this paper is to place a yellow-green emitting $SrBaSiO_4:Eu^{2+}$ phosphor layer or a red-emitting $Sr_wF_xB_yO_z:Eu^{2+},Sm^{2+}$ phosphor layer on the $YAG:Ce^{3+}$ phosphor layer and then select the most appropriate value of $Sr_wF_xB_yO_z:Eu^{2+},Sm^{2+}$ concentration to get the greatest color quality. Experimental results were exactly as expected since $Sr_wF_xB_yO_z:Eu^{2+},Sm^{2+}$ really succeeded in increasing CRI and CQS. Specifically, the larger the $Sr_wF_xB_yO_z:Eu^{2+},Sm^{2+}$ concentration, the higher the CRI and CQS, resulting from the increase of the red light component in WLEDs. Meanwhile, the quality degree of luminous flux depends tightly on the $SrBaSiO_4:Eu^{2+}$ phosphor layer. However, if the $Sr_wF_xB_yO_z:Eu^{2+},Sm^{2+}$ and $SrBaSiO_4:Eu^{2+}$ concentration exceeds, the luminous flux and color quality will have a tendency to drop off, according to the theory of Mie scattering and the Lambert-Beer law. The result of this paper is a valuable reference in the production of high-quality WLEDs.

1. Introduction

Nowadays, as the demand for lighting is increasing and complex, traditional light sources become obsolete and need to be replaced. In that situation, phosphor converted white light emitting diodes (pc-WLEDs), the fourth potential generation light source, was invented to dominate the illumination market and become widespread in different fields of our daily life such as landscape, street lighting, backlighting, etc [1, 2]. However, there is still a limitation in light extraction efficiency and angular homogeneity of correlated color temperature which obstructs the white LED in developing its integrity. Therefore, in order to expand the consumption market as well as optimize the use of white LEDs, these two aforementioned shortcomings need to be removed. Moreover, combining the blue light from converse red phosphor with the yellow light from LED chip is considered as the optimal method for the white light generation [3].

The first idea focused on the use of the $SrBaSiO_4:Eu^{2+}$ green phosphor [4] layer to increase the green light component in WLEDs, resulting in increased photon emission. In contrast, in the second idea, the red phosphoric layer $Sr_wF_xB_yO_z:Eu^{2+},Sm^{2+}$ was used to increase the red light component in WLEDs, thereby increasing CRI and CQS. The article also details the chemical composition of $Sr_wF_xB_yO_z:Eu^{2+},Sm^{2+}$ and their effects on the optical properties of WLEDs. The article's results show that CRI and CQS are increased with the addition of the $Sr_wF_xB_yO_z:Eu^{2+},Sm^{2+}$ phosphor layer. However, an appropriate concentration of $SrBaSiO_4:Eu^{2+}$ and $Sr_wF_xB_yO_z:Eu^{2+},Sm^{2+}$ should be selected in order to avoid a sharp drop in color quality or the output luminous flux, as the concentration of green or red phosphors increases. There are two noticeable differences when the green or red phosphor layer is added into the yellow $YAG:Ce^{3+}$ phosphor layer. First, the rise in green or red light leads to the increase in white light spectra, which is also the key to advance color quality. Second, the scattering and light transmission in WLEDs are in the direction of the concentration of additional phosphors. Therefore, the selection of suitable phosphor concentration becomes essential for the maintenance of photosensitivity of WLEDs.

2. Construction of green-yellow dual-layer phosphor configuration and red-yellow dual-layer phosphor configuration

Figure 1 (a) showed the picture of WLEDs with 9 internal LED chips which is used in this study. The output of each blue chip is 1.16W with the peak value emitted at 453 nm. The detail of LED profile specifications was illustrated in Figure 1 (b).

Lead frame: 4.7 mm Jentech Size-S

LED chip: V45H

Die attach: Sumitomo 1295SA

Gold Wire: 1.0 mil

Phosphor: ITC NYAG4_EL

(a) (b)

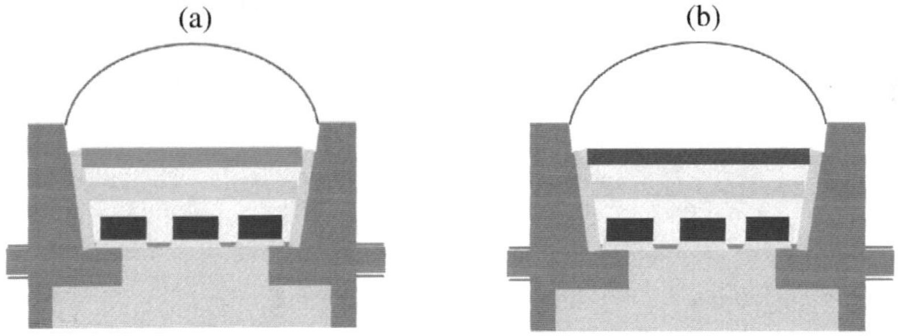

(c) (d)

Figure 1. Illustration of pcWLEDs: (a) The actual WLEDs and (b) its parameters; (c) Illustration of GYC, and (d) RYC.

To find the optimal concentration of $SrBaSiO_4:Eu^{2+}$ and $Sr_wF_xB_yO_z:Eu^{2+},Sm^{2+}$, the next step is to construct the remote phosphor model. In this study, two models of green-yellow dual-layer phosphor configuration (GYC) and red-yellow dual-layer phosphor configuration (RYC) were proposed. The GYC structure consists of two phosphor layers placed on the blue chips. $SrBaSiO4:Eu^{2+}$ phosphor layer is located above the yellow phosphide layer $YAG:Ce^{3+}$, see Figure 1 (c). In Figure 1 (d), the RYC structure includes two phosphor layers putting on the blue chips. $Sr_wF_xB_yO_z:Eu^{2+},Sm^{2+}$ phosphor layer is placed above the yellow $YAG:Ce^{3+}$ phosphor layer. The goal of using GYC and RYC configurations is to improve the color and the output luminous flux of WLEDs by increasing the scattering and composition of both green light and red light in WLEDs. However, the schema of $SrBaSiO4:Eu^{2+}$ and $Sr_wF_xB_yO_z:Eu^{2+},Sm^{2+}$ needs to be calibrated accordantly.

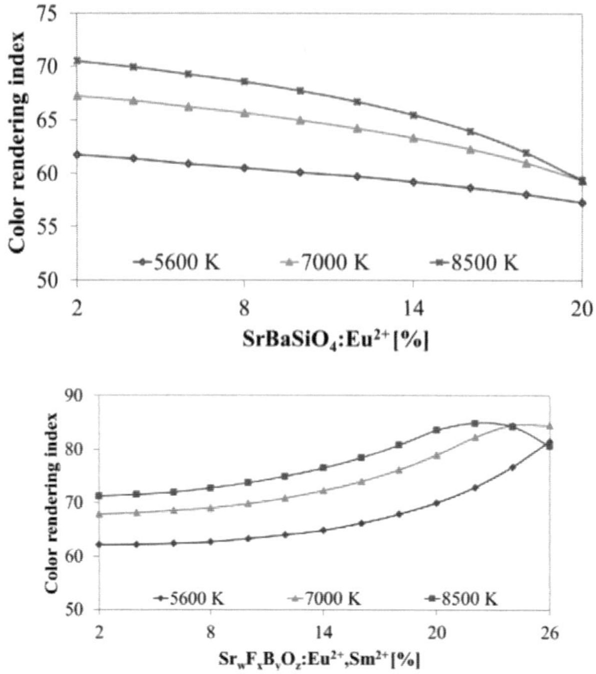

Figure 2. The color rendering index as a function of the concentration of $SrBaSiO_4:Eu^{2+}$ and $Sr_wF_xB_yO_z:Eu^{2+},Sm^{2+}$: (above) GYC; (below) RYC.

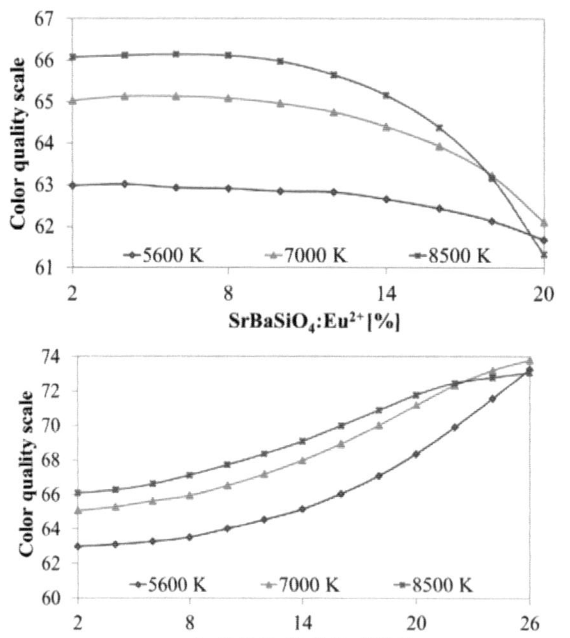

Figure 3. The color quality scale as a function of the concentration of $SrBaSiO_4:Eu^{2+}$ and $Sr_wF_xB_yO_z:Eu^{2+},Sm^{2+}$: (above) GYC; (below) RYC.

IOP Conf. Series: Materials Science and Engineering **617** (2019) 012014 doi:10.1088/1757-899X/617/1/012014

3. Computation and discussion

In this article, the color rendering index is used to evaluate the true color of the object when the light is shining. The amount of green light goes up, causing the imbalance among three main colors: blue, yellow, and green. Consequently, the color quality of WLEDs is affected, resulting in the decrease of color accuracy of WLEDs. The results in Figure 2 (above) show the reduction of CRI in the presence of the $SrBaSiO_4:Eu^{2+}$ remote phosphor layer. Nevertheless, this is still acceptable because compared to CRI, CQS is more vital and difficult to achieve. In Figure 3 (above), CQS does not change when the $SrBaSiO_4:Eu^{2+}$ concentration is less than 8%. Hence, 8% $SrBaSiO_4:Eu^{2+}$ is an ideal concentration when this phosphor is applied in WLEDs manufacturing after considering the emission of the flux.

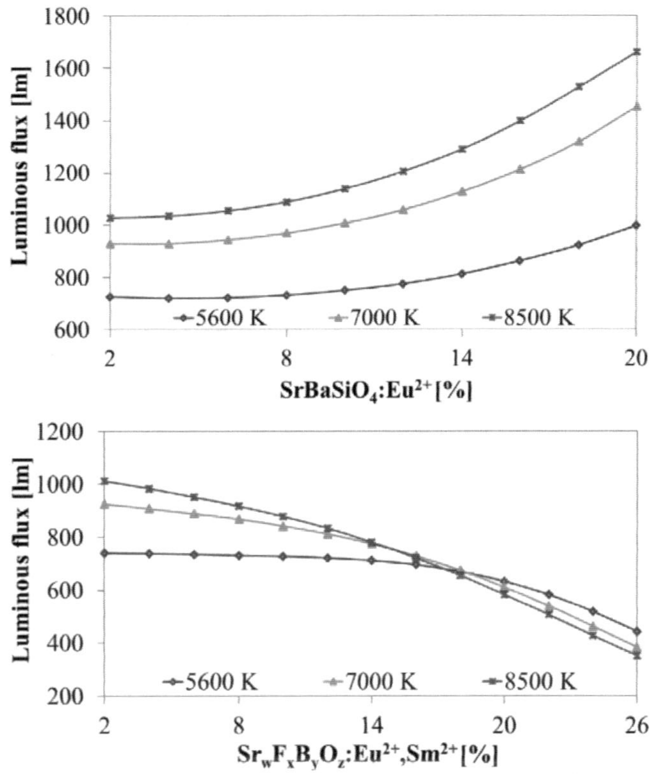

Figure 4. The luminous flux as a function of the concentration of $SrBaSiO_4:Eu^{2+}$ and $Sr_wF_xB_yO_z:Eu^{2+},Sm^{2+}$: (above) GYC; (below) RYC.

As can be seen from Figure 2 (below), the colorimetric index increases with the $Sr_wF_xB_yO_z:Eu^{2+},Sm^{2+}$ phosphor concentration at all average CCTs. This is the result of the absorption of red phosphor. When $Sr_wF_xB_yO_z:Eu^{2+},Sm^{2+}$ phosphor absorbs blue light from the LED chip, the red phosphor particles convert the blue light into red light. Beside the blue light from the LED chip, $Sr_wF_xB_yO_z:Eu^{2+},Sm^{2+}$ particles also absorb the yellow light. However, in these two absorptions, the blue light absorbed by the LED chip is stronger due to the absorption properties of the material. Therefore, the red light component in WLEDs increases with the addition of $Sr_wF_xB_yO_z:Eu^{2+},Sm^{2+}$, resulting in a higher color rendering index (CRI). In the current WLED selection, the color rendering index is one of the important parameters. Of course, the higher the color rendering index, the higher the price of white light WLED. However, if the $Sr_wF_xB_yO_z:Eu^{2+},Sm^{2+}$ phosphor is used to produce WLEDs, the cost will be reduced considerably. As a result, $Sr_wF_xB_yO_z:Eu^{2+},Sm^{2+}$ can be widely applied to this manufacturing process. However, the color quality of WLEDs cannot be evaluated by only CRI. A WLED with good color quality does not mean it has a high color rendering index too. Therefore, this study provides a new parameter called color quality ratio (CQS). CQS is an index defined by three factors: the color index, the viewer's preference, and the color coordinates. With these three vital factors, CQS becomes a powerful index that can measure the overall color quality of WLEDs. Figure 3 (below) shows the elevation of CQS with the presence of the remote phosphor layer $Sr_wF_xB_yO_z:Eu^{2+},Sm^{2+}$ and the sharp increase of CQS when the concentration of $Sr_wF_xB_yO_z:Eu^{2+},Sm^{2+}$ appears. Obviously, using the $Sr_wF_xB_yO_z:Eu^{2+},Sm^{2+}$ phosphor can improve the white light quality of WLEDs with dual-layer phosphor structures. This important result nearly reaches the aim of improving color quality of the study. However, the disadvantage of $Sr_wF_xB_yO_z:Eu^{2+},Sm^{2+}$ to the output luminous flux cannot be overlooked. Figure 4 (above) showed that the resulting emission increased dramatically when $SrBaSiO_4:Eu^{2+}$ concentration increased from 2% wt. to 20% wt. However, the dual-layer remote phosphor door is affected by the concentration of the $Sr_wF_xB_yO_z:Eu^{2+},Sm^{2+}$ phosphor layer. Noticeably, according to the Lambert-Beer law, the attenuation coefficient µext is proportional to the concentration $Sr_wF_xB_yO_z:Eu^{2+},Sm^{2+}$ except the ratio of light to energy. Therefore, when fixing the thickness of both phosphor layers of WLEDs, the photon emission can be reduced as $Sr_wF_xB_yO_z:Eu^{2+},Sm^{2+}$ concentration increases. As predicted, Figure 4 (below) showed the decrease in overall three CCTs. When the concentration of $Sr_wF_xB_yO_z:Eu^{2+},Sm^{2+}$ is at 26%, wt., the luminous flux decreased extensively. However, if considering the advantages of the $Sr_wF_xB_yO_z:Eu^{2+},Sm^{2+}$ phosphor layer including improved CRI and CQS, in addition, the flux of the dual-layer remote phosphor is also higher than the single-layer phosphor layer (without the red phosphor layer), this decline is completely acceptable. The remaining problem depends on the manufacturer's intention to select an appropriate concentration of $Sr_wF_xB_yO_z:Eu^{2+},Sm^{2+}$ when mass production of this WELDs is conducted.

4. Conclusion

In conclusion, this paper presents the effects of $SrBaSiO_4:Eu^{2+}$ phosphor and $Sr_wF_xB_yO_z:Eu^{2+},Sm^{2+}$ on CRI, CQS, and photon of dual-layer phosphor structure. Based on Mie's scattering theory and the Lambert-Beer law, the study has proven that $Sr_wF_xB_yO_z:Eu^{2+},Sm^{2+}$ is the right choice for bettering the color quality. Meanwhile, $SrBaSiO_4:Eu^{2+}$ is the choice for innovating WLEDs. This is not only applied for WLEDs with a low color temperature of 5600 K but also for those with color temperatures above 8500 K. Therefore, the results of this research have achieved the goal of improving the color quality of white light, which is very difficult for the remote-phosphor structure. However, there is still a slight disadvantage to the emission of the flux. If the increase of the $SrBaSiO_4:Eu^{2+}$ or $Sr_wF_xB_yO_z:Eu^{2+},Sm^{2+}$ levels is rapid, the color quality or luminous flux will decrease sharply. Consequently, selecting a suitable concentration becomes more important than ever. Depending on the manufacturer's purpose, the decision will be made to produce better quality WLEDs based on the information from the reference this article provided.

References

[1] B. –Y. Joo and J. –H. Ko, 2013 "Analysis of Color Uniformity of White LED Lens Packages for Direct-lit LCD Backlight Applications," *Journal of the Optical Society of Korea*, vol. **17**, pp. 506-512.

[2] W. -S. Sun, C. –L. Tien, J. –W. Pan, T. –H. Yang, C. –H. Tsuei, and Y. –H. Huang, 2013, "Simulation and Comparison of the Lighting Efficiency for Household Illumination with LEDs and Fluorescent Lamps," *Journal of the Optical Society of Korea*, vol. **17**, pp. 376-383.

[3] M. Cantore, N. Pfaff, R. M. Farrell, J. S. Speck, S. Nakamura and S. P. Denbaars, 2016, "High luminous flux from single crystal phosphor-converted laser-based white lighting system," *Optics Express* vol. **24**, pp. A215-A211.

[4] W. M. Yen and M. J. Weber, Inorganic Phosphors: Compositions, Preparation and Optical Properties, (CRC Press, LLC, 2000 N.W. Corporate Blvd., Boca Raton, Florida 33431, 2004).

IOP Conf. Series: Materials Science and Engineering **617** (2019) 012015 doi:10.1088/1757-899X/617/1/012015

MOCVD growth GaN on sapphire

O Abdullaev[1], M Mezhenny[1], A Chelny[1], A Savchuk[1,2], Yu Ahmerov[1], O Rabinovich[2], V Murashev[2], S Didenko[2], Yu Osipov[2], S Sizov[2], M Orlova[2] A Aluyev[1]

1 JSC "Optron", 105187, Moscow, Russia
2 National university of science and technology "MISiS" 119049, Leninskiy pr., bl.4, Moscow, Russia
E-mail: rawork2008@mail.ru

Abstract. The n-type and p-type a-GaN films were successfully grown on a r-sapphire substrate, according to X-ray diffractometer and SEM results parameters measurement. The growth rate versus the growth temperature was investigated. The holes concentration ($8x10^{17}$ cm^{-3}) was achieved by the Cp_2Mg flow optimization and the parameters of thermal annealing in nitrogen. The GaN film growth rate dependence versus temperature at a constant hydrogen flow through a TEG source was investigated. The results indicate that defects density is reduced upto 10^4 cm^{-2}, the surface morphology uniformity was improved. During growth the influence from V/III flows ratio was detected.

1. Introduction

InGaN/GaN films growth in nonpolar a-orientation is an important scientific and technological goal. It is known that the polarization effect leads to GaN bulk and surface changes in the band structure. The built-in polarization electric field leads to the LEDs efficiency reduce due to the quantum-dimensional Stark effect [1-7]. In quantum wells, carrier current separation occurs, the degree of wave functions overlapping reduces and the injection efficiency decreases. The built-in polarization field magnitude increases with indium concentration rise in the InGaN/GaN. This leads to a decrease in injection efficiency with a wavelength increase for LEDs and lasers. The proper solution is the semi-polar and non-polar substrates usage, for which the polarization field in the growth direction is minimal or completely absent. The grown films will reduce the effect of the built-in field on the injected charge carriers distribution, will reduce the nonradiative recombination rate and improve the efficiency of the emitting devices.

However, there were some difficulties in achieving the comparable with c-plane results. Mostly, such difficulties are due to high dislocation density [8], to different impurities behavior in c-plane material [9] and another epitaxial growth kinetics [10]. One of the most promising and prospective way to grow high quality nonpolar GaN films is r-sapphire substrate usage [11, 12]. In this work this approach was used to a-GaN films growth. A complex structure investigation by X-ray diffraction, SEM Raman scattering spectra and SIMS allows to enlarge understanding of nonpolar GaN growing process.

2. The experimental procedure

GaN has a hexagonal wurtzite lattice and belongs to the space group C^4_{6v}. The atoms dislocation corresponds to the C_{3v} symmetry. GaN films were grown by MOCVD on r-sapphire substrates $(11\bar{2}0)$ under reduced pressure (80 torr) in a horizontal reactor with preliminary surface nitridization. The sources for III and V components were triethylgallium (TEG) and ammonia (NH_3), respectively. The layered gas flow in the quasi-laminar regime in the reactor was carried out through a porous input. The upper gas mixture flow consisted of N_2 and NH_3 and the lower flow - H_2 and TEG. As sources of donor and acceptor impurities, hydrogen sulphide (H_2S) (0.5% mixture in argon) and biscyclopentadienyl magnesium (Cp_2Mg) were used, respectively. Preliminary, the sapphire substrate surface was nitrided at a temperature 950 °C for 10 minutes. For single crystal a-GaN growth MOCVD parameters were optimized. At temperatures above 870 °C, the film became a single crystal due to the fact that in this temperature range the GaN growth rate did not depend on temperature, which was common to ensure the layer thickness reproducibility of the material being grown. At lower temperatures, the kinetic nature of the surface process takes place, and the gallium nitride growth rate is characterized by such temperature dependence. V/III ratio was in range 200 - 1000 for different samples. The sample structure is presented in the Table 1.

Table 1. InGaN structure.

0.3 [μcm], $Al_{0.1}Ga_{0.9}N$
0.4 [μcm], p-type GaN
32 [nm], QW
0.4 [μcm], n-type GaN
0.4 [μcm], AlGaN
1.9 [μcm], $Al_{0.1}Ga_{0.9}N$

3. Results and discussion

In Figure 1 it is shown an image of the a-GaN / r-Al_2O_3 structure based on SEM. As it can be seen in Figure 1a the main defect type are triangle pits, oriented in c-direction. Such defects size is no more than 1 mcm, and concentration of such defects is 10^4 cm^{-2}. On the cross-section view these defects can be observed as triangle pyramids. The structure of cross section SEM view is familiar with previously achieved results on r-sapphide [13].

Figure 1. a- grown GaN film surface; b - cross section of the sample.

The film growth rate versus temperature dependence on a constant TEG flow rate is shown in Figure 2.

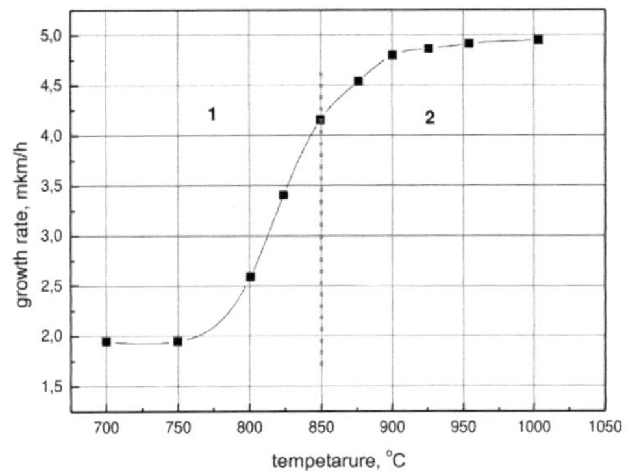

Figure 2. GaN film growth rate versus temperature at constant flow rate TEG $f = 400\ cm^3\ /\ min$: 1) amorphous crystalline phase; 2) crystalline phase.

The diffraction patterns from a two-crystal X-ray diffractometer are presented in Figure 3. It can be seen that the grown film corresponds to a-GaN.

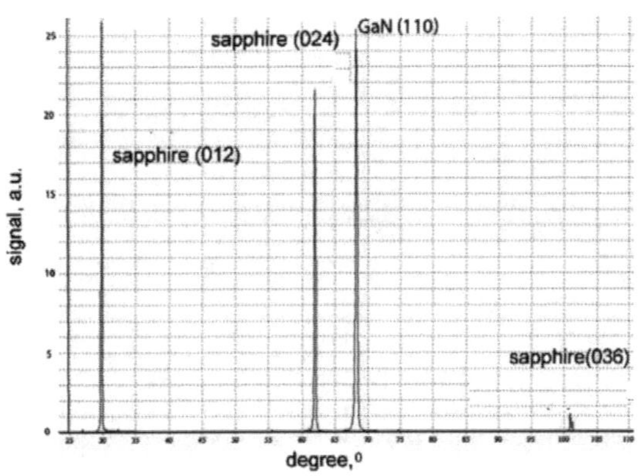

Figure 3. Diffraction pattern of a-GaN film on r-Al_2O_3 substrate.

X-ray diffraction patterns of gallium nitride films on sapphire are shown in Figure 4.

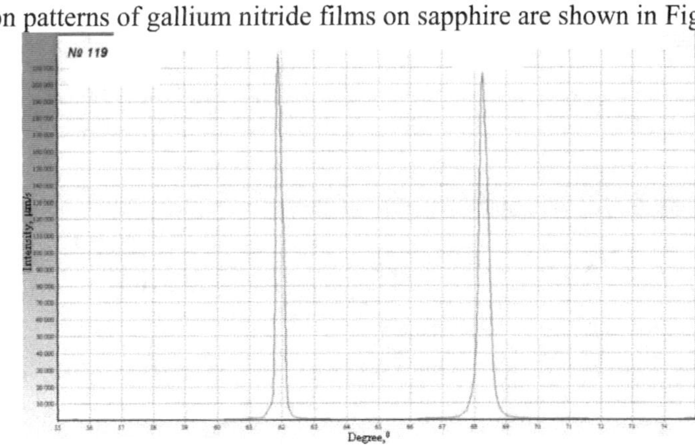

Figure 4. X-ray spectra of deposited GaN films on the r-sapphire substrate and the sapphire itself.

At temperatures above 870 °C, the film became single crystal, the best morphology was achieved at a temperature - 900-920 °C. From the obtained diffraction patterns, it is seen that single-crystal area films have been grown on the r-substrate. The Raman scattering spectra is presented in Figure 5.

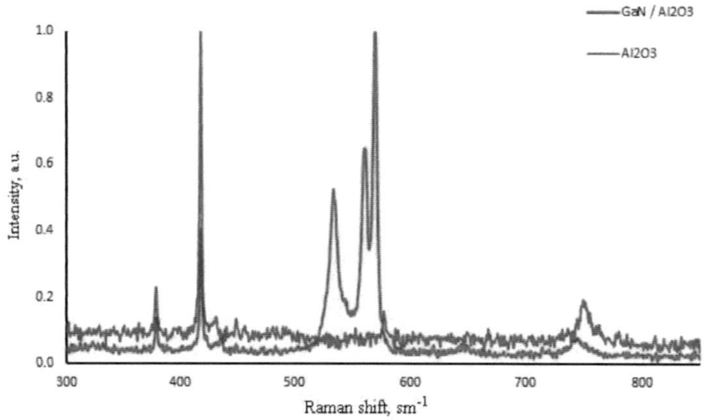

Figure 5. Raman scattering spectra of deposited GaN films on the r-sapphire substrate and the sapphire itself.

It is seen that the Raman spectrum of the Al_2O_3 – r-sapphire substrate is characterized by peaks - 379, 418, 578, 751 cm^{-1}. For both GaN films types a triple Raman peaks with maxima about 534, 561, 570 and 742 cm^{-1} are detected.
The carrier concentration dependences versus the film growth parameters are shown in Figure 6.

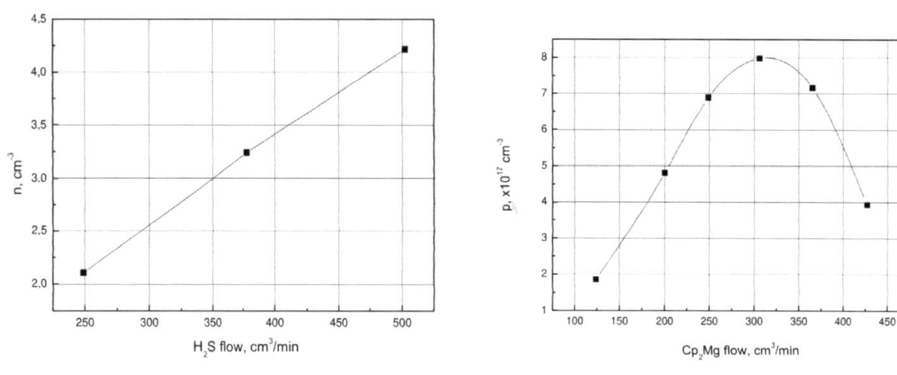

Figure 6. Carrier concentration versus the film growth parameters.

It is detected that with doping for n-type the dependence is linear, for p-type the extremum is detected. The obtained spectra for the GaN films under investigation, can be correlated with a gallium nitride film.
The measured half-width of the films rocking curve is 1.5-2 times larger than for films with "polar orientation". This is due to the different structure defects density in GaN layers with different orientations, namely nanopipes, packing defects, screw, edge and mixed dislocations [14-15]. With film thicknesses rise, the crystal perfection increases, which is due to the decrease edge dislocations influence. Based on the etch pits shape, the dominant dislocations types are mixed dislocations.
The luminescence spectrum at room temperature is shown in Figure 7.

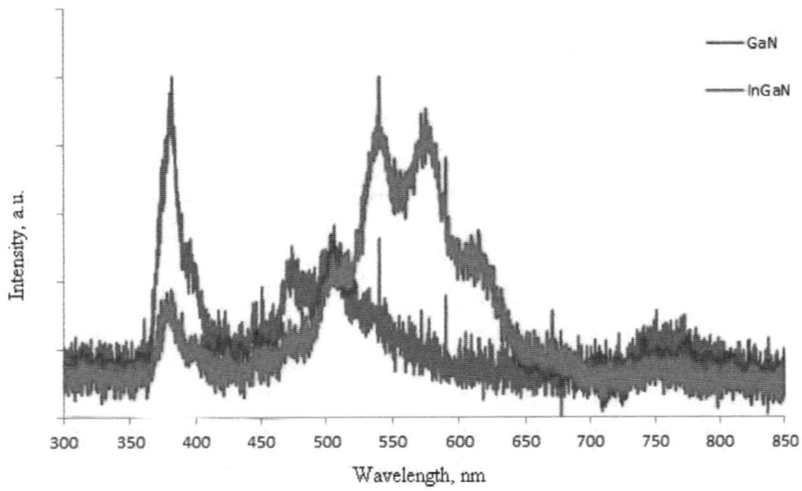

Figure 7. GaN-r film PL on a sapphire substrate.

A clear peak is observed at 370 nm, which corresponds to the GaN band-band transition. In the graph the yellow and green luminescence peaks are clearly seen. These peaks are explained by recombination between the levels created by the defects complexes. The following defects are known, for which the observed luminescence characteristic is: V_{Ga}, V_{Ga}-On and Cn Cn-On complexes.
It was developed the technology for InGaN growth on the synthesized material. The In concentration measurements by SIMS are shown in Figure 8.

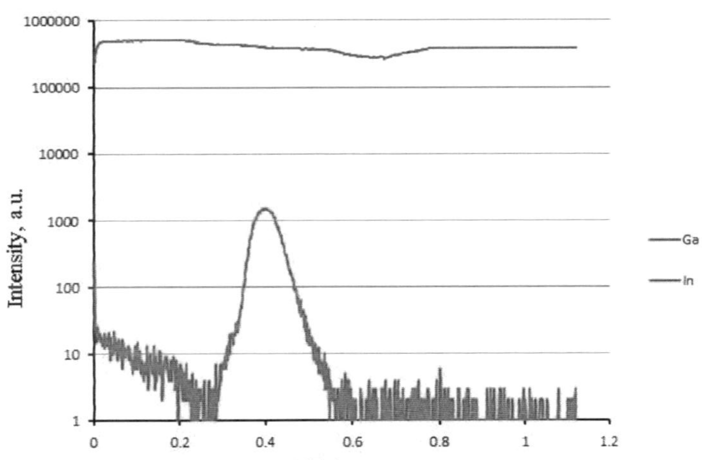

Figure 8. InGaN film elements distribution vs thickness.

4. Conclusions

By the original method, a-GaN was successfully grown on r-sapphire by MOCVD. Using the X-ray and Raman scattering spectrum, it was confirmed that the grown films are a-GaN single-crystal films. The dominant defects type on the film surface were investigated by SEM: the defects similar by nature to V-defects on the c-plane. This defect type has been investigated for non-polar orientation films for the first time.
Decrease in the V / III ratio upto 950 at the stage of the precipitated growing low-temperature nuclei leads to the growth islands formation, and it also increases the growth rate in the lateral direction. This leads to a decrease in the dislocations density. a-GaN films on the r-Al_2O_3 substrate were grown by the

MOCVD with good surface morphology. The acceptor and donor impurities concentrations were $8 \cdot 10^{17}$ cm^{-3} and $4 \cdot 10^{18}$ cm^{-3}, correspondently.

References

[1] Aderhold J., Davydov V.Yu. and Fedler F. 2001 *J. of Cryst. Growth* **222** 701.
[2] Strite S. and Morkoc H. 1992 *J. Vac. Sci. Technol. B* **10** 1237.
[3] Nakamura Sh., Senoh M. and Nagahama S. 1996 *Japanese Journal of Applied Physics* **35** L74.
[4] Fedorchenko I.V., Kushkov A.R., Gaev D.S., Rabinovich O.I., Marenkin S.F., Didenko S.I., Legotin S.A., Orlova M.N. and Krasnov A.A. 2018 *J. of Cryst. Growth* **483** 245.
[5] Fedorchenko I.V., Ril A.I., Marenkin S.F., Rabinovich O.I., Legotin S.A. and Didenko S.I. 2017 *J. of Cryst. Growth* **468C** 683.
[6] Rabinovich O.I., Didenko S.I., Legotin S.A., Fedorchenko I.V. and Osipov U.V. 2015 *Journal of nano- and electronic physics* **7(4)** 040351.
[7] Rabinovich O., Legotin S., Didenko S., Yakimov E., Osipov Yu. and Fedorchenko I. 2016 *Japanese Journal of Applied Physics* **55** 05FJ131.
[8] Liliental-Weber Z., Jasinski J. and Zakharov D.N. 2004 *Opto-electronics review* **12(4)** 339.
[9] Brownea D.A., Young E.C. and Lang J.R. 2012 *J. of Vacuum Science & Technology A* **30** 4.
[10] Lymperakis L. and Neugebauer B. 2009 *J. Phys. Rev. B* **79** 241308(R).
[11] Wu Z.H. and Fischer A.M. 2008 *Appl. Phys. Lett.* **92** 171904.
[12] Craven M.D., Lim S.H. and Wu F. 2002 *Appl. Phys. Lett.* **81** 469.
[13] Wu F., Craven M.D. and Lim S.-H. 2003 *J. of Appl. Phys.* **94** 942.
[14] Jinno D., Otsuki Sh. and Niimi T. 2017 *Physica status solidi (b)* **254** I8.
[15] Wang C., Jiang Y. and Die J. 2019 *Cryst. Eng. Comm.* **21** 2747-2753.

IOP Conf. Series: Materials Science and Engineering **617** (2019) 012016 doi:10.1088/1757-899X/617/1/012016

Heritage Building Era Detection using CNN

Md. Samaun Hasan[1], Dr. Mohammad Ali[2], Mizanur Rahman[3], Hosne Ara Arju[4], Mirza Mohtashim Alam[5], Md. Salah Uddin[6], Dr. Shaikh Muhammad Allayear[7]

1,3,5,6,7 Department of Multimedia and Creative Technology, Daffodil International University, Dhaka, Bangladesh.
2 Department of Graphic Design, Crafts & History of Art, Rajshahi University, Rajshahi, Bangladesh
4 Department of Sanskrit, Rajshahi University, Rajshahi, Bangladesh
Email: salah.mct@diu.edu.bd

Abstract. The Indian subcontinent is a southern area of Asia continent which includes India, Bangladesh, Pakistan, Nepal, Bhutan, Maldives, and Sri Lanka. In the different periods, different rulers had ruled in these territories such as the Sultanate period (1206–1526) and Mughal period (1526–1540, 1555–1857). In addition, various ancient and heritage structure patterns show the different historical and religious characteristics of the old civilizations. This research presents a computational method for identifying the construction period or era of old heritage building by using feature detection. We have done this experiment for the application aspect, where the tour guide is not available this application will help the tourists for knowing the construction period or era by detecting the features of old spectacular architecture. In this study, we have focused on the constructional characteristics of old architectural sites by using Canny Edge Detector method. After using the Canny Edge Detector method, we have categorized different types of features among the buildings such as Dome, Minaret or Tower, Ratna, and Front. The different eras old structure contains different characteristics of the Dome, Minaret or Tower, Ratna, and Front. Finally, we have developed a neural network based model which can identify the building's construction era of the Sultanate period and Mughal period.

1. Introduction

Detection of an object is currently an important research area in the field of computer vision and artificial intelligence. Building Detection is one of them [1]. The Indian subcontinent is a natural terrain in South Asia. There are many old tourist sites in this landmass such as Sona Mosque, Kusumba Mosque, Sixty Dome Mosque (Bangladesh), Taj Mahal, Qutub Minar (India), Tomb of Jahangir (Pakistan) etc. Day by day, many unknown and unexplored old structures are coming to the people's eye. For that, travelers are excited about seeing the recent revealed ancient structure [2]. Sometimes, foreign tourists cannot find a knowledgeable tour guide [3]. In this situation, it is hard for tourists for acquiring the knowledge of construction period of ancient buildings, mosque, temple, etc. In this point of view, we have explored this research where a software application helps the tourists knowing the construction period or era by detecting the old spectacular architecture. In recent years, some research papers have been published, where machine learning and computer vision approaches are used in ancient architecture and archaeology section [4]. A feature recognition technique is used in ancient architecture which is based on deep learning and computer vision [5]. In this paper, the researchers have proposed a method to identify the several features of the ancient structure by using an artificial neural network. Another recent method focuses on recognition and visualization for ancient Maya hieroglyph where the researchers utilized CNN technique [6]. Furthermore, these former researchers have not expressed any method for identifying the construction period or era of ancient architectural structure like the ancient temple, mosque, and building. Due to this research gap, we have committed this research where an intelligent technique helps the tourists for knowing the construction period by detecting the old spectacular architecture.

2. Proposed system and process

In this research, presents an inquiry, field observation and an analytical study about the building detection of different eras. From many sources, we get the information of some mosques, temples, tombs, bridge, dispensary, schools, the palace of Zamindars, and the houses of general people. Some antiquities are fully destroyed and some are partly destroyed. This research will suggest taking an initiative of the preservation of all ancient structures due to insufficient resources.

Our research has illustrated an application based model which demonstrated how a computer program can identify the era of an ancient building. At first, an image was taken and sent to the Canny Edge Detector. After applying this method we have got two different types of dataset: a training set and test set. Then we created a CNN model to get the prediction result of old eras (Sultanate period and Mughal period). The steps and architecture of the proposed system have been shown in Figure 1.

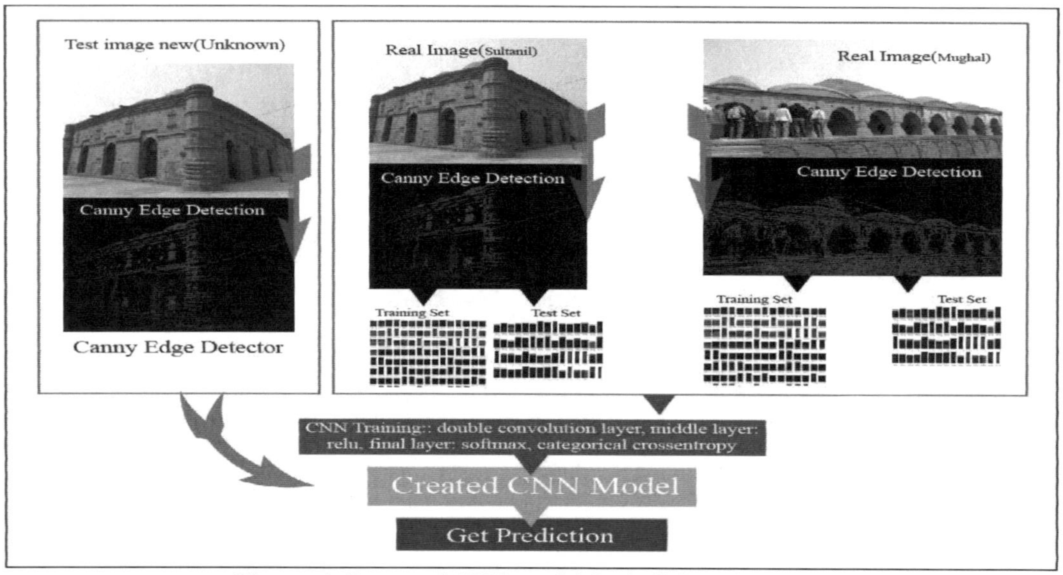

Figure. 1. Proposed CNN model for building detection.

3. Building structure and feature identification

There are some computational methods for detecting the features and characteristics of buildings. Research has been narrated for identifying the damaged segment of a historical building [7]. In image processing perspectives, there are different types of pattern for feature recognition [8], such as edges, corners, points, etc. [9]. In our experiment, we have used some techniques for collecting the building features of the Sultanate and Mughal period.

The outcomes of the proposed model construct of realizing the old era and a machine provides probable data by learning the ancient building's features. We have denoted how a programmable machine theoretically learns the different ancient buildings' features such as Dome, Minaret or Tower, Ratna, and Front. By using the Canny Edge Detector method we got different Dome, Minaret or Tower, Ratna and Front structure of the Mughal and Sultanate architecture. The learning process is based on these characteristic structures. Figures 2-15 respectively, have been exhibited the feature detection of the Sultanate and Mughal period architectures for identifying the construction era. The training set was based on those characteristics and detected features. For detecting the edge, we used the Canny Edge Detector because this function has been utilized for detecting a wide range of edges from the image. Numerous researches agree with the Canny approach showing the best result [10].

IOP Conf. Series: Materials Science and Engineering **617** (2019) 012016 doi:10.1088/1757-899X/617/1/012016

The basic difference between two eras (Sultanate Period and Mughal Period) is its structure, decoration, and materials. In the Sultanate Mosques of Bengal have a unique exposition of surface decoration including stone-cut design, terracotta, color marble, and pleasuring ornamentation, but the Mughal mosques are simply ornamented with plaster in different styles such as a rectangle, arched panel, flat, four-centered or multi-foiled arches. Due to lack of easy availability of architectural material like a stone within the short distance, the Mughal mosque continued to be dominated by brick masonry in its building art and the molded terracotta made out of the fine-textured alluvial clay for purposes of surface decoration. The plastered wall is some of the typical features of the Mughal architecture of Bengal.

Figure. 2. Image of sultanate period.

Figure. 3. Image after using canny detection of sultanate period.

Figure. 4. Front image of sultanate period.

Figure. 5. Front image after using canny detection of sultanate period.

Figure. 6. Minaret image of sultanate period.

Figure. 7. Minaret image after using canny detection of sultanate period.

Figure. 8. Front image of mughal period.

Figure. 9. Front image after using canny detection of mughal period.

IOP Conf. Series: Materials Science and Engineering **617** (2019) 012016 doi:10.1088/1757-899X/617/1/012016

Figure. 10. Ratna image of mughal period.

Figure. 11. Ratna image after using canny detection of mughal period.

Figure. 12. Minaret image of mughal period.

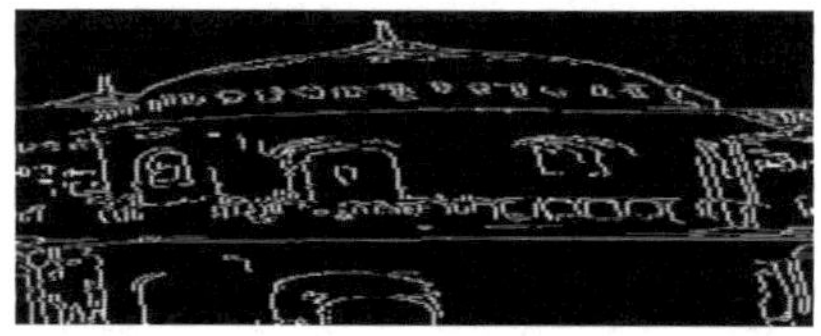

Figure. 13. Minaret image after using canny detection of mughal period.

Figure. 14. Dome image of mughal period.

Figure. 15. Dome image after using canny detection of mughal period.

4. Proposed network architecture

We have created a Convolution neural network using 3 convolution layers 2 max pooling layers and 2 fully connected layers and a dropout (see Figure 16). Input image with size 64*64 and channel one is inserted into the first convolution layer.

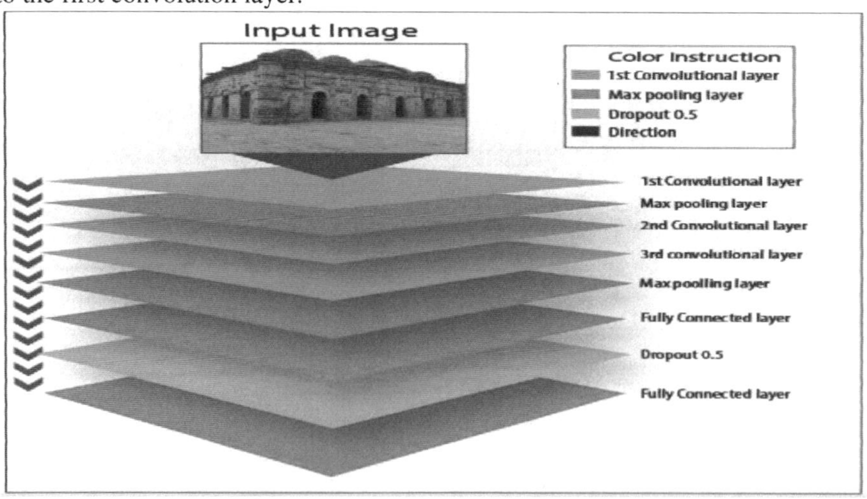

Figure. 16. Diagram of CNN model.

5. Experimental result

The outcomes of the proposed model construct of realizing the old era where a machine provides data by learning the ancient building's features.

Confusion matrixes mainly describe the performance of a classifier model and it contains information about actual and predicted classifications which are done by a classification system [11]. For evaluating the performance of such systems, we used the data in the matrix. We trained our CNN model with our modified dataset and calculated the accuracy.

Test Dataset 1: Firstly, we made a single prediction in our CNN model using a single image which successfully predicted buildings of the Sultanate (see Figure 17) and Mughal (see Figure 18) period. Finally, we got a 100% accuracy rate for a single image.

Name	Type	Size	Value
edges	uint8	(800, 1200)	array([[0, 0, 0, ..., 0, 0, 0], [0, 0, 0, ..., 0, 0, 0],
i	int	1	1
img	uint8	(800, 1200, 3)	array([[[210, 186, 144], [213, 189, 147],
prediction	str	1	Sultani
result	float32	(1, 2)	array([[0., 1.]], dtype=float32)
test_image	float32	(1, 64, 64, 3)	array([[[[0., 0., 0.], [0., 0., 0.],

Figure. 17. Predicted result of Sultanate period

Name	Type	Size	Value
edges	uint8	(800, 1200)	array([[0, 0, 0, ..., 0, 0, 0], [0, 0, 0, ..., 0, 0, 0],
i	int	1	1
img	uint8	(800, 1200, 3)	array([[[210, 186, 144], [213, 189, 147],
prediction	str	1	Mughal
result	float32	(1, 2)	array([[1., 0.]], dtype=float32)
test_image	float32	(1, 64, 64, 3)	array([[[[0., 0., 0.], [0., 0., 0.],

Figure. 18. Predicted result of Mughal period.

Test Dataset 2: Total tested data 600. The Sultanate building contains data 350 & Mughal building 250 (see Table 1).

Table 1. Confusion matrix of raw data model.

SULTANATE	337	13
MUGHAL	33	217

From Table 1, we get TP = 337, TN = 217, FP = 33, FN = 13 and we know that,

$$\text{Accuracy} = \frac{(TP+TN)}{(TP+FP+TN+FN)} \times 100\% \tag{1}$$

Following above formula (Eq. 3) for the raw data, we have achieved 92.33% accuracy.

Test Dataset 3: Total tested data 400. The Mughal building contains data 130 & Sultanate building 270.

Table 2. Confusion matrix of raw data model.

SULTANATE	254	16
MUGHAL	3	127

From Table 2, we get TP = 254, TN = 127, FP = 3, FN =16 and we know that,
Following above formula (Eq. 3) for the raw data we have achieved 95.25% accuracy.

6. Conclusion and future work

Our research has illustrated an application based model which demonstrates how a computer program can identify the era from an ancient building. This research is mainly focused on the period of building identification by using feature detection. Though we have achieved much better accuracy still there are somewhat limitations that we have put aside for our future work. Further, if we test the model with blurry or distortion picture, it cannot determine the target result of that specific picture. This drawback would also lead us to our future work to make our model more robust and more significant to recognize precise objects from the image. These issues will be looked forward to solve in proper research.

7. References

[1] Moya L. et al., "Detection of collapsed buildings from lidar data due to the 2016 Kumamoto earthquake in Japan," Nat. Hazards Earth Syst. Sci., vol. 18, pp. 65–78, 2018.
[2] Unexplored tourism destinations of India, FICCI and EY, Ernst & Young LLP, 2015.
[3] Chockalingam M. and Ganesh A. A., "Problems encountered by tourists," Business and Economic Horizons, vol. 3, no. 3, pp. 68-72, 2010.
[4] Hörr C., Lindinger E., Brunnett G., "Machine learning based typology development in archaeology," ACM Journal on Computing and Cultural Heritage, vol. 7, no. 1, article 2, 2014.
[5] Zou Z. et al., "Feature recognition and detection for ancient architecture based on machine vision," Proc. SPIE 10602, Smart Structures and NDE for Industry 4.0, 1060209, United States, 2018.
[6] Can G., Odobez J., Gatica-Perez D., "How to Tell Ancient Signs Apart? Recognizing and Visualizing Maya Glyphs with CNNs," ACM J. Comput. Cult. Herit., vol. 9, no. 4, article 1, 2018.
[7] Crespo C. at al., "Damage detection on historical buildings using unsupervised classification techniques," International Archives of Photogrammetry, Remote Sensing and Spatial Information Sciences, vol. 24, XXXVIII, Part 5, 2010.
[8] Ghosh P., Pandey A. and Pati U. C., "Comparison of Different Feature Detection Techniques for Image Mosaicing," ACCENTS Transactions on Image Processing and Computer Vision, vol. 1, no. 1, pp. 1–7, 2015.
[9] Thareja M. and Goyal A., "Performance Analysis of Edges, Corners and the genres: A Subjective Estimation," IOSR Journal of Electronics and Communication Engineering, vol. 1, Conf.15010, pp. 98–104, 2016
[10] Maini R. and Aggarwal H., "Study and Comparison of Various Image Edge Detection Techniques," International Journal of Image Processing, vol. 3, no.1, pp. 1–11, 2009.
[11] W. Zhang, G. Zelinsky, and D. Samaras. Real-time accurate object detection using multiple resolutions. In Proc. IEEE International Conference on Computer Vision, 2007.